The Four Keys to Successful Design

The Four Keys to Successful Design

✦

A motivational approach, from thought to finish.

Nathan Todd Cool

iUniverse, Inc.
New York Lincoln Shanghai

The Four Keys to Successful Design
A motivational approach, from thought to finish.

iUniverse, Inc.

For information address:
iUniverse, Inc.
2021 Pine Lake Road, Suite 100
Lincoln, NE 68512
www.iuniverse.com

ISBN: 0-595-29909-1 (pbk)
ISBN: 0-595-66077-0 (cloth)

Printed in the United States of America

This book is dedicated to the person in my life who, from my youngest days, has brought me inspiration, sparked my innovation, and helped me to explore the things I can create: my brother, Dan Cool.

Contents

Key #4: Creation

Introduction

If you chose a career to design, engineer, architect, or invent, this book was written with you in mind. If you've ever been curious about how things work, wanted to make things better, and wanted to turn your ideas into realities, then you and I have a lot in common.

Ever since I can remember, I've always wondered how things worked. Everything puzzled me. How did those pictures appear on the TV? Why did sound come out of a stereo? What made airplanes fly?

Curiosity is what led me to the field of engineering. When I was young, I envisioned the field of technology and science as a noble profession in which I would not only satisfy my curiosity, but also build great things. Technology and science inspired me, and set the course for my career path.

Once deeply involved in the science of engineering, finding answers to why things work, and looking at a bright future of making the world a better place, I quickly learned that engineering could be a very dull, process-oriented profession. Project planning, budgets, schedules, risk assessment, resource allocations, spreadsheets, and technical documents were not exactly my idea of changing the world with new technical marvels. Still, this never squelched my desire to delve into the unknown, solve its problems, and create that which others desired.

I found, through the quagmire of engineering bureaucracy, that key fundamentals for creative design often go unnoticed and overlooked. Without proper mentoring, many engineers never fully develop these essentials. These missing pieces of the puzzle are not just simple rudiments of constant value, like many principles we are taught in school. Instead, they are deeply rooted foundations from which you, as a creative person, can unlock further potential.

In particular, I found four key fundamentals that are common to many successful designs, not just from my own experience, but from well-known innovators, engineers, and designers as well. I choose to call these four key fundamentals "The Four Keys to Successful Design." These keys are not just steps in a process.

Instead, they are pearls of wisdom that unlock our potential to engineer, design, and build successful products.

Inspiration is the first of the four keys that we'll learn to master in this book. In engineering-oriented careers, it's not always easy to break through the hard crusty surface of official procedure surrounding the sweet juicy center of scientific and creative designs. Salivating for this core substance drives our desire to be, and to do. This desire is called inspiration, a fundamental disregarded through many scholastic curriculums, but one that we'll study closely to understand and strengthen.

Innovation, the second of the four keys, requires the key of Inspiration to take us through a sometimes difficult journey of perspective thought, which often encounters speed bumps and roadblocks along the way. Nevertheless, mastering these first two critical keys of Inspiration and Innovation help us to break through our obstacles and come out on top. We're then ready to take on the exciting third key of Exploration, leading us to the challenging fourth and final key of Creation, where we see our original thoughts finally turn into realities.

Inspiration, Innovation, Exploration, and Creation are the four keys that unlock our inner beings, bring us into the world of wonder, and lead us through the challenge of technical pursuit. This is what makes us *successful* engineers, designers, architects, and inventors, and not just everyday cube-dwelling 9-to-5'ers performing the mundane tasks of procedural development. Inspiration, Innovation, Exploration, and Creation are not just elements from which you can build a process. Instead, these symbiotic entities can enhance your abilities to create, invent, and deliver no matter what the product of your goal may be.

Engineering is more than the stereotypical geekdom of paint-by-numbers design. It's more than just a job. It's more than just sitting in a gray-walled cube pounding away on the keyboard in front of you. Engineering is being creative, seeing things from a different angle, exploring the unknowns in science, and bringing ideas to realization. This holds true no matter what part you play in the game. From drafters to architects, from programmers to aerospace engineers, from clothing designers to toy makers, we all can utilize the four keys presented in this book to enhance our potential for successful design.

Engineers are not just bridge builders or skyscraper architects. To be an engineer is to be someone with an idea, and the ability to take that idea through to completion. You could be a web designer building sites on the Internet. You could be a process engineer improving methods for product development. Or you could be a biologist in search of better explorative devices. The list goes on and on. The title of Engineer goes a long way, and covers a lot of ground. So do the four keys presented in this book.

These four keys are woven into many standard engineering processes used today. Yet these keys also dig deeper into the heart and soul of our creative thought and desire for progress. Inspiration, Innovation, Exploration, and Creation are the keys that delve further into our engineering essence, and gain momentum in our forward drive to successful designs.

I have found that these four keys help not only in my engineering career but in my personal life as well. The four keys are not entirely bound to the process of engineering, but the nature of who we are, and go beyond the realm of that gray-walled cube. The four keys reach out into the world in which we exist. Similar to laws of nature, the four keys are naturally occurring elements. But just like with gravity, inertia, and relativity, we need to understand the four keys, and utilize and manipulate them to our advantage during the engineering process.

I want to share these keys with you, and discuss the lessons I've learned from them. I want to present these keys to you from not only my perspective, but also through the eyes of some of the great people throughout history that have brought me inspiration. Their influence has given me motivation that kept me going through the toughest of technological and bureaucratic challenges, as well as personal hardships. As such, you will find a variety of quotations and examples throughout this book, referencing those things that helped lay the foundation for the four keys.

So what exactly are these four keys, and what can you expect from them? Here's a quick look at what we'll be covering:

- **Inspiration**: We'll take a look at why this is more than a key, and such a vital element for successful designs. We'll see how the key of Inspiration is glue, bonding the other three keys together. As you'll see throughout this book, without this primary component the symbiotic relationship between all four keys will break down. We'll take a closer look at the nature of inspiration; how to find it, understand it, and defeat its worst enemies.

- **Innovation**: What is innovation? Where does it come from? We'll discuss how this key plays the role of visionary among the other three keys. We'll look at how this key does more than just foster ideas. This key will show you how innovation requires unique perspectives as we journey through all four keys.

- **Exploration**: This key goes beyond fact-finding, and hunts out faultfinding as well. We'll also see how its lower predecessors, Inspiration and Innovation, fuel this key. We'll examine how to avoid pitfalls, and how to pick ourselves up by the bootstraps when we inevitably hit brick walls and speed bumps along the way.

- **Creation**: During this key, we'll take a brief look at some standard engineering processes, and how they are influenced by all four keys. We'll look at maintaining a positive course of Inspiration, Innovation, and Exploration throughout this last key with efficient, practical approaches that lead us to our final product: a successful design.

After covering the four keys, we'll sum it all up by hitting the highlights of these keys in the Summation section. I've provided this section as a synopsis, a quick and easy read that briefly touches the topics previously discussed. This will give you a quick review of the aspects that stand out most among the four keys.

So without further ado, let's delve deeper into these four keys, and unlock your potential for successful design.

Key #1: Inspiration

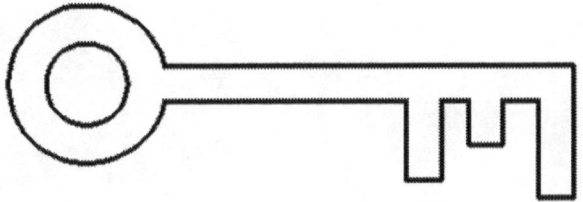

1

Inspirational Soul Food

*Take the first step in faith. You don't have to see the whole staircase,
just take the first step.*

—*Dr. Martin Luther King Jr.*

The *four keys* presented in this book—Inspiration, Innovation, Exploration, and Creation—are comparable to a food chain. Bigger fish eat little fish. At the lowest end of this chain is the littlest fish, Inspiration. A bigger fish known as Innovation feeds on Inspiration to nourish itself. And even further up this chain is Exploration, which requires a steady diet of innovation to sustain its course into the phase of Creation. At each step along the way, if we break the lowest link in this chain by removing inspiration, the system will collapse.

This is where it all begins: being inspired by something that drives you to the next step. A quick thought passes through your mind. You visualize a new or better way of doing something. An idea hits you like a brick. A light comes on in your head. An epiphany strikes you when you least expect it. Something spawned this train of thought. Something called inspiration.

As engineers, designers, architects, and inventors, we need inspiration to drive our innovative ideas. But did any of us take "Inspiration 101" in college? Why didn't they teach us the most basic fundamental of engineering and creative design? Most often, when it comes to inspiration, it's taken for granted that either you have it or you don't. But the fact is that we all have it. However, we need to take a close look at it, understand it, nourish it, and help it to grow.

Where does inspiration come from, and how do we give it strength? The answers are closer than you think. They are right inside of you and ready to come out.

3

Inspiration takes many forms and shows up in some of the most unexpected places.

The story of Velcro®, that amazing material that's used in countless applications today, is a story of inspiration that lead to innovation. In 1951, Swiss inventor Georges de Mestral went on a walk with his dog. Upon returning home, he noticed that his dog's coat and his pants were covered with nasty little cockleburs. Those annoying burrs were difficult to remove from both the dog and his pants. Instead of getting frustrated from those sticky little burrs, Georges was inspired.

Those annoying cockleburs had amazing "sticking" power, and Georges wanted to know why. If those little burrs could cling to material so strongly, perhaps he could design something based on these bothersome little thistles for practical use. Could something be artificially made to simulate the power of these remarkable little cockleburs?

Georges closely studied the burrs under a microscope, where he discovered a difference in the burrs as compared with other nettles and weeds. The cockleburs had a natural hook-like shape, allowing them to grab hold of material instead of just poking it. Thus was born the innovation for a unique, two-sided fastener—one side with stiff "hooks" like the burrs, and the other side with the soft "loops" like the fabric of his pants.

The result was VELCRO® brand hook and loop fasteners, named for the French words "velour" (the soft, looped side) and "crochet" (the hooked side). If it not for Georges de Mestral's inspiration from the amazing sticking power of cockleburs, a great innovation may never have been born. It was the key of inspiration that led the way, and innovation came as a result.

We don't often perceive inspiration as an engineering principle, and it slips through the cracks of learning. Whenever we hear of popular designs, breakthrough technologies, and cutting-edge products, we often hear the term "innovation." It was "an innovative design" or it was "an innovative breakthrough." Nevertheless, the four keys, like any food chain, have a symbiotic and dependant relationship that creates a careful balancing act amongst them: Inspiration, Innovation, Exploration, and Creation. Without some kind of inspiration, innovation would simply starve to death. Without inspiration, Georges de Mestral would have merely grumbled with frustration as he removed the bothersome burrs from his dog and pants.

Note however that having only the thoughts and questions as to why, how, and what is not exactly inspiration. We can think of inquisitiveness as the cornerstone of inspiration, but it takes more than just curiosity to cultivate inspiration to feed innovative thought.

Imagine yourself in a field of grass, the sun shining brightly above you, clouds passing quietly overhead as you ponder the nature of the things around you. The year is 1490 A.D., and so much is unknown about the world in which you exist. You might wonder such things as why the sky is blue. What created those clouds? Why is the wind blowing from the south? Then something puzzles you even more. A flock of birds flies overhead. You think to yourself, "How do they do that? Why am I, as a human, grounded to earth? Yet birds can travel freely through the air without the constraints of land and its various obstacles?" You have questions and wonder. But is this inspiration?

Think about how many people throughout history asked the same question: "How do birds fly?" Then compare that to how many people attempted to create transportation that could fly (like airplanes and helicopters). Many people possess wonder, but fewer get inspiration from their curiosity. Therein lies the difference between inquisitiveness and inspiration, a dichotomy that separates the innovative from the merely curious.

The mystery of flight inspired Leonardo Da Vinci. It drove him through a study of the science with such works as his "Flying Machine" in 1490, his "Flexible Wing" design in 1495, and many other designs he drafted circa 1500. Although Da Vinci never created a successful flying machine, his inspiration was remarkable. He did not give up, and over the span of more than a decade, he pursued the possibility of human flight. Although Da Vinci's inspiration of flying machines never went as far as a successful innovative design, it inspired him to continue designing and served as an inspiration to many engineers and artists that followed throughout history.

One could conclude that inspiration is truly inherent in only a select few people with innate talents: the Da Vincis, the Wright brothers, the Thomas Edisons, and the Einsteins. After all, these people were special. Were they not the geniuses of their times (even by today's standards)? True, these were exceptional people. Nevertheless, we all have something in common with the great inventors, visionaries, and innovators throughout history. We all have the gift of thought and the potential for innovation driven by inspirational strength.

All of us have inspiration. If it were not for inspiration, we would go through life without trying even the simplest of things. Have you ever tried to cook, play soccer, play an instrument, or better yet, study engineering at a university? All of these took inspiration and the courage to advance that inspiration towards the next step. If you think to yourself that you are not inspired, I'd lay odds that you are wrong. Inspiration comes in many forms; it's a matter of finding yours.

Inspiration, like our personalities, is unique to our individual selves. This gives each of us varying appetites for this "soul food"—the first source of engineering sustenance that feeds the creative mind. Throughout our lives, each of us experiences different moments in time, different upbringings, different schooling, and other various factors. These factors form and mold us into unique persons, each with a distinct way of thinking, problem solving, and reasoning. Because of this uniqueness, each of us has a different level of inspirational hunger. Some of us have insatiable appetites no matter how sour, or sweet, each bite is. On the other hand, some of us lose our appetites upon the first sniff of slightly tainted fruit.

How would you rate your inspirational appetite? How inspired are you? The following chapters regarding the first key of Inspiration will delve deeper into those questions, pointing out some fundamentals that can help build your inspiration.

Inspiration is the initial stride in a journey through engineering, architecture, and creative design. Inspiration is a vital key that we cannot overlook. It is critical that we nurture this gentle seed, and learn to foster an environment in which it can grow and blossom within us, thus allowing our creations to be formed from truly *inspired* innovative thought. It is for this reason that we need to accept inspiration as a foundation upon which we can build the other three keys.

Let's now wet our appetites and increase our hunger for the staple of our creative engineering diet: inspiration. The remaining chapters in this section regarding inspiration will help you to take a closer look inside yourself, and better understand the first of *The Four Keys to Successful Design*: Inspiration.

2

Successes and Failures

Success is the ability to go from one failure to another with no loss of enthusiasm.

—*Sir Winston Churchill*

To grasp a full understanding of inspiration, we need to face its toughest nemesis: failure. Like a military general preparing for battle, we need to know our enemy well. In the case of this enemy, however, our objective is not to kill it. Instead, our goal is to welcome it, take it in, and use it to our advantage.

If at first you don't succeed, it makes it tough to keep going. But as the saying goes, we have to "try and try again." That's easier said than done. After all, we are humans, not machines, and as such we have feelings and emotions. We tend to be encouraged and complimented by our successes, not our failures. When we took our first steps, said our first words, or got A's on our first report cards, some form of compliment typically followed these events. Each time there was a success, encouragement (in most cases) followed.

Inspiration thrives on past successes and encouragement. If you solve a problem to a physics equation, you feel confident and inspired to tackle another similar or more difficult challenge. If you're playing basketball with friends, and make a half court shot, camaraderie runs high and compliments abound (except perhaps from the opposing team).

Failure, on the other hand, is the antithesis of success. What if you can't solve that physics problem, or you never score in your game? Human nature tends to read failure as a negative thing, which depletes our motivation and starves our inspirational appetite. Not solving a problem or having poor "game" can drastically reduce your inspiration. Nevertheless, you can manipulate these negatives, turn failure into a useful tool for positive success, and increase your inspiration.

It's uninspiring to fail. If you have a great idea but it just doesn't work out, you might see this as defeat. Or perhaps others scoffed at your idea, making it seem like a total waste of your time. These events can be uninspiring. Failure is not something we tend to thrive on. So should we avoid it, thus feeding inspiration with nothing but accomplishments? That's not only impossible but impractical as well. If we never met failure, we'd lose some valuable lessons and have fewer experiences to build on. By understanding human nature and having confidence in ourselves, we can turn lemons into lemonade. We don't have to avoid failure or be afraid of it. Instead, we need to welcome it and learn from it.

The invention of 3M's Post-it® Notes is a story of failure. Sound bizarre? How could such a successful product be a failure? Actually, the Post-it® Notes story is one of failure that turned to success, and inspiration lead the way from start to finish.

If you are not readily familiar with the Post-it® Notes, they are those little self-stick notepapers that typically come in yellow pads (although there are many more colors today). Most people have, use, and love Post-it® Notes. But how did they come to be?

A man named Spencer Silver was working in the 3M research laboratories in 1970 to develop a new strong adhesive. Silver developed a new adhesive, but it was even weaker than what 3M already made. It stuck to objects, but could easily be lifted off. It was super weak instead of super strong. The design backfired. This sticky substance was practically the opposite of what Silver set out to create.

Nobody at 3M knew what to do with the stuff. Nevertheless, Silver did not throw it away or write off the whole project as a failure. Silver used this as experience, and moved on. Most importantly, he didn't throw away the adhesive and its formula.

Instead, Silver knew that he had invented a highly unusual new adhesive. He didn't look at this as a failure; he saw it as potential success, and a learning experience to recall for future exploration.

Now the challenge Silver faced was what to do with the weak sticky substance. For the next five years, Silver gave seminars and buttonholed individual 3M'ers, extolling the potential of this new adhesive and showing samples of it in spray-can form and as a bulletin board. Nevertheless, this idea never really took hold. The product continued looking like a failure. But time would prove otherwise.

Shortly afterward, another 3M scientist, Arthur Fry, who had attended one of Silver's seminars on the wimpy adhesive, was singing in the church's choir. He used paper bookmarkers to keep his place in the hymnal, but these pieces of paper kept falling out of the book. Remembering Silver's adhesive, Fry used some of the weak goo to coat his bookmarkers. Viola! With the weak adhesive, the markers stayed in place, yet lifted off without damaging the pages. This super weak adhesive, which originally was a failure, had a purpose, and was now on the road to success.

But the story isn't over yet. The sticky-edged paper faced more obstacles ahead. Skeptics remained within 3M as attempts were made to launch this new product. In many aspects, it seemed as though this idea was still a failure. Engineering and production people told Fry that Post-it® Notes would pose considerable processing measurement and coating difficulties, and would create a lot of waste. Market research was a tough nut to crack as well. Who would pay for a product that seemed to be competing with cost-free scrap paper? Despite the obstacles that made Post-it® Notes seem like a failure, 3M went forward and marketed the product. In 1981, one year after its introduction, Post-it® Notes were named the company's Outstanding New Product.

3M, originally known as the "Minnesota Mining and Manufacturing Company," created a culture in which designers and inventors aren't criticized for their failures. Later on in Chapter 3, "The Company You Keep," we'll be taking a close look at how companies that foster a motivational environment maintain your inspiration. For now, I'd like to briefly touch on this subject to point out how important 3M's culture was for the inventors, making it easier for Silver and Fry to learn success from failure with continued enthusiasm and inspiration.

A man named William L. McKnight joined 3M in 1907 as an assistant bookkeeper. He quickly rose through the ranks, becoming president in 1929 and chairman of the board in 1949. Many believe McKnight's greatest contribution was as a business philosopher, since he created a corporate culture that:

1. Encourages employee initiative

2. Encourages employee innovation

3.   Provides secure employment.

These three elements foster a culture where failure is not looked upon as an evil entity.

McKnight's basic rule of management was laid out in 1948, which states:

> "As our business grows, it becomes increasingly necessary to delegate responsibility and to encourage men and women to exercise their initiative. This requires considerable tolerance. Those men and women to whom we delegate authority and responsibility, if they are good people, are going to want to do their jobs in their own way.

> "Mistakes will be made. But if a person is essentially right, the mistakes he or she makes are not as serious in the long run as the mistakes [that] management will make if it undertakes to tell those in authority exactly how they must do their jobs.

> "Management that is destructively critical when mistakes are made kills initiative. And it's essential that we have many people with initiative if we are to continue to grow.

The last paragraph in McKnight's statement really hits home when it comes to failures turning into inspiration. McKnight realized the importance of failure, and not to be afraid of it. To this day, 3M's culture fosters creativity and gives their employees the freedom to take risks, accept failure, and try new ideas.

But what if you don't work for 3M and have such a great motivational force behind you? How does one take a failure and turn it into something positive to stay inspired? In short, it's a matter of perspective. Successes naturally grow inspiration. When we succeed, we see the fruits of our labor as bright, shining, and beautiful. Hence, we accept our successes as good things, which in turn drive our inspiration.

On the other hand, with failure we tend to look at the fruits of our labor as though they are rotted, dull, and ugly. We turn away from these failures and abandon them, sometimes seeing them as appalling reflections of ourselves. But this is a false perception of reality, and failure is not an ugly thing. As we abandon our failures, we inadvertently leave behind any trace of inspiration that could be lying deep within that not-so-pretty fruit.

Instead of seeing failure as hideous, repulsive, distorted pictures of what we pre-fer, take a moment and look at it another way. See the beauty that lies deep within that failure. Accept the failure, analyze it, and use it to your advantage. Remember that failures are foundations upon which you can build better things. Experience can be the hardest substance known to man, and the best thing to hold together the future of your designs.

We'll be discussing failures a bit more throughout the four keys. We'll take a look at other failures and obstacles, and I'll show you ways to manage these as well. For now, realize that failure is not all bad. Stay inspired knowing that deep within each failure lies a success waiting to happen. As in the case of the Post-it® Notes story, this could take years. Nevertheless, the benefits remain just the same.

3

The Company You Keep

On the next leg of our journey through the first of the four keys, we take a stop at a critical juncture that influences your inspiration: your environment.

I often pose a question to myself when sitting at my job, cornered into a square workspace with cloth walls, no ceiling, and privacy fit for a caged monkey. Who in their right mind would want to be an engineer? In this field, the majority of companies will stuff you in a small cubicle, place you in front of a computer, and surround you by the artificial glow of fluorescent lighting. What kind of a think-tank is that? Oftentimes, sitting in my gray, drab cube, I ask myself the same question. Nevertheless, there is a darn good answer why. It is the company I keep or, looking at it from another perspective, the company that is keeping me. It all depends on how you look at it. In either case, working for the right employer, even if that means yourself, plays a major role in building your inspirational spirit.

If it is just a job, well then, I guess things aren't all that bad sitting in the "cube farm" day in and day out, enjoying the drab grey fabric walls encompassing your dimly lit domain of labor. However, what does it take to inspire someone who wants to be creative, grow, foster his or her inspirational appetite, and pursue the sciences of engineering?

If you want to strengthen your inspiration, and motivate innovative thought, an important fact you need to address is the selection of the company you keep, or the company that is keeping you. Whether you are working for a company or self-employed, your work and work environment play a vital role in fostering your inspirational spirit.

Sure, you could put a fern in your workspace, and perhaps pictures of the kids, some cool calendars, and a motivational poster or two (depending on how tiny that cube may be). But these are only facades, merely icing on a much thicker cake. Real inspiration comes from deep within, and so does the substance of your working environment and employee satisfaction.

Many companies today realize that their employees can be more productive by not slave-driving them but instead fostering an inspirational environment. Computer World Inc., whose web site is at http://www.computerworld.com, has a special report entitled "The Best Places to Work in IT," located at http://www.computerworld.com/bestplaces2002/.

This report, besides listing some great companies to work for in the IT field, also sheds light on employee satisfaction for other types of jobs. The report is a compilation of survey data taken from CEOs, VPs, and Directors at U.S.-based companies. The survey asked about such things as training, development, average salary increases, percentage of staff promoted, turnover rates, mentoring programs, and benefits.

Employee satisfaction is a hard thing to measure, but the Computer World report seems to have hit the mark. Many of the surveys taken in the industry focus on numbers but don't dig deeper into the motivation behind the answers. The Computer World report, however, has a number of articles that go beyond the numbers gathered from the top brass surveyed and delve deep into the "whys." Here are a few of the highlights from the Computer World report that explain why some companies maintain high employee satisfaction:

1. Mentoring, formal training, and career development planning. As this report and many other articles regarding employee satisfaction point out, it is not just the money that employees are after. Working for a company that cares about your career future is a win-win situation. The company wins with better-educated and experienced staff. The employee wins with gratification and pride, both of which offer great nourishment for inspiration.

2. Offer the hottest technology to work on. Why did you get into a career in engineering, design, or architecture anyway? Most of us can answer that by simply saying we want to learn more, feeding our curiosity about what makes things tick, etc. So why work on outdated technology, or products that cannot motivate you? Let's quickly revisit Nora Roberts' quote from the beginning of this chapter. Two things in that quote drive home this important concept. First, "If you don't go after what you want, you'll never have it," and second "If you don't step forward, you're always in the same place." Do not be afraid to move forward, if that is what you want. Do not get caught in the past if this would squelch your desire to progress and hinder your inspiration.

3. Strike a balance between work and life. It's difficult to be inspired by your work if you're stressed out from the daily activities of your home life. After all, do we live to work, or do we work to live? There is a balance between the two scenarios. Do not let your work consume you. Remember there is another world outside the confines of your cube, and so much more that this world has to offer besides just work.

This year, Rice University conducted an employee satisfaction survey of their Facilities & Engineering (F&E) team. This, as with many such surveys available for public access, points out some other fundamentals for satisfying an employee. Here's a few of the high-ranking items that made the list:

- My work matters.
- I get to do my best work.
- I have opportunities.
- There is consistent teamwork.
- My boss cares about me.

Can you relate to these? Think about how much time you spend at work. If you spend eight hours a day in the office, then you are talking about 50 percent of your waking hours during the week. That's half of your life (at least from Monday through Friday) spent in the environment of your employer.

If you feel your work doesn't matter, that it is just a waste of time, that no one will use what you are working on, then inspiration will be hard to come by. However, if you get to perform at your best level, and your employer allows you to expand into other opportunities, then your work does matter, and inspiration is sure to follow.

Camaraderie is paramount. A positive teamwork environment is essential to staying motivated and inspired. As human beings, we need the interaction of others. This gives us positive reinforcement, provides guidance from peers, and gives us the benefits of collective thought. Positive results are inherent to a good team. Inspiration is built upon the whole, not just the individual parts.

If you are working for a company that is not allowing your inspirational spirit to flourish, then it may be time to take a step forward. You may need to look towards a future that can provide encouragement and satisfaction in your work. You are an engineer, a designer, an architect, or an inventor. If that title means little to you, then I would like to point you to something that Herbert Hoover was quoted as saying:

> *"Engineering is a great profession. There is the fascination of watching a figment of the imagination emerge through the aid of science to a plan on paper. Then it moves to realization in stone or metal or energy. Then it brings homes to men or women. Then it elevates the standard of living and adds to the comforts of life. This is the engineer's high privilege."*

Obviously stated before the age of computers and other technologies we have today, this quote holds true for all facets of engineering. It's a great profession to turn logical thought into computer programs, to create communication equipment that brings our world closer together, to design affordable housing for the masses, to make efficient warm clothing materials, grow healthier and safer foods, and partake in many more areas in which engineering plays a role in the lives of our populous planet.

If Hoover's quote moves you, and you realize that you deserve respect by not only your peers but also by the employer for which you toil, then tear off that pocket protector, throw down your adhesive-taped, black-framed glasses, jump on top of your desk, and yell above all the cubicles in the room, "I am an engineer! I am a designer! I am a creator of great things!" If that should get you fired or labeled as a lunatic, I hereby disclaim any and all responsibility for your actions. On second thought, perhaps you should just sit back quietly, smile, and bask in the warmth of knowing you are worthy and noble. Heck, even Herbert Hoover said so.

Aristotle said it best when he stated:

Pleasure in the job puts perfection in the work

How true that is. Quality companies have quality people who make quality product. The reason for this is simple. Foster a creative and inspirational environment, and innovations will be right around the corner. When people are happy with their work, they find new and better ways of performing it. This holds true for every brick in the human resource structure composing a company, from the janitor to the CEO, and every engineer, designer, and architect in between. If we find pleasure in our work, we will be inspired, and great things will come as a result.

One last point I would like to address concerning the "company that is keeping you" involves the morality of your employer. While most companies flourish based on their credibility, a few out there do not always play by the rules. Without boring you with details about my experience working for one such unscrupulous company that I walked away from, let me just say that the reputation of the company is not the only thing at stake, it is *your* reputation as well. What is most important to you? Companies will come and go. Employers will come and go. But you have to live with yourself for the rest of your life. To stay inspired, I suggest you sleep well at night knowing you are working for a credible employer. Knowing each day that you are not working in a reputable environment can be depressing, and will inevitably hinder your inspiration. Don't be afraid to move forward if your heart and morality are telling you that it's time.

Feel free to choose your employer, and not just let them choose you. Never allow anyone, including your employer, to dry up your sea of desire, leaving you with thirst and no inspiration. Doing so is robbing your soul of its deepest desires. Inspiration is far too valuable to waste.

I'd like to close out this chapter with a few more quotes to drive home how important your work is to your inspirational spirit. I hope you find these as meaningful as I have:

Derive happiness in oneself from a good day's work, from illuminating the fog that surrounds us.—Henri Matisse

Get happiness out of your work or you may never know what happiness is.—Elbert Hubbard

Getting fired is nature's way of telling you that you had the wrong job in the first place.—Hal Lancaster, in The Wall Street Journal

Nothing is really work unless you would rather be doing something else.—James M. Barrie

In order that people may be happy in their work, these three things are needed: They must be fit for it: They must not do too much of it: And they must have a sense of success in it.—John Ruskin

Never continue in a job you don't enjoy. If you're happy in what you're doing, you'll like yourself, you'll have inner peace. And if you have that, along with physical health, you will have had more success than you could possibly have imagined.—Johnny Carson

4

Introspection

Keep away from people who try to belittle your ambitions. Small people always do that, but the really great ones make you feel that you too, can become great.

—Mark Twain

So far, we have talked about the periphery of Inspiration. We have seen how inspiration thrives on accomplishment, and how we can manipulate failure to do the same. We've explored what it takes to ensure that our work environments are best suited to motivate our inspirational appetites as well. These principles are paramount in nurturing our inspiration, and maintaining it through the remainder of the four keys. However, we've only been discussing outside influences that effect our inspiration. Before moving on to the remaining three keys of Innovation, Exploration, and Creation, we need to get to the core of inspiration located deep within ourselves.

As we move forward through this book, you will see how intertwined all four keys are, and how inspiration plays a vital role each step of the way. I cannot stress enough how important it is that you feel inspired, and find ways to continually feed your inspirational appetite, yet keep hungering for more. Inspiration is more than just a spark that ignites innovation. It's also the glue that bonds together the other three keys. Inspiration is a critical element that, if lost, will cause the entire structure of our innovations to collapse.

Inspiration is something we all possess in varying degrees. It's not reserved for the elite, it knows no bounds, and it is well within your reach. At this point in the book, you might already feel inspired. Perhaps you can look back at people in your life that influenced you, and helped to establish the basis for your inspirational drive towards engineering and creative design. However, if you find inspiration difficult to muster up, it doesn't mean you lack it or that you never will have it. It simply means it may take a little more work for you to strengthen this vital key.

In either case, in this last chapter on the first of the four keys, I would like to take the time to point out some fundamental principles of Inspiration, and how each of us can find our inspiration and work to reinforce it.

Directing our inspirational spirits towards careers in engineering, design, or architecture takes reflection and introspection. We need to look inside ourselves, examine how we feel about what we're doing, and realize our strengths, potentials, and true desires. We need to ask ourselves some difficult questions and answer them honestly. If we don't, we will only be lying to ourselves and muting our inspirational potential. Ask yourself this:

- What is it that I truly want to do in life? Is it really designing, engineering, creating, and developing things? If not, ask yourself why you are pursuing this career path. Is it because someone else suggested it, or is it income-related? If you are sure that this career is the right thing for you, then you can build your inspirational spirit as you continue to pursue your goals and dreams. If not, then you may never be truly inspired; your true inner being would not be satisfied.

- What is it that I do well? Engineering and design encompass a broad field. It can vary from writing software, designing cars, architecting houses, designing clothes, creating video games, developing medical lifesaving equipment, and the list goes on. Realize the field of engineering science (or creative design) you wish to pursue, and be true to yourself in your selection.

Before answering those questions, step back for a moment and look deep inside yourself. What are your strengths and potentials, the things that will drive your inspiration? What motivates you? What brings you true happiness? It's time to look in the mirror, and see what reflects back. We need to see inside ourselves and ask these pointed questions with introspection. The following are just a few points[1] that can help give you insight into this. Stop for a moment and ponder these things with true introspection:

1. Look at your yearnings, perhaps activities you have always been drawn to since childhood, even if you've never had the time or opportunity to express them. Don't look at these things as childhood fantasies. Look at these as possible avenues for your career path. Why go through life and never pursue your dreams? Work is not all about money; it's about wealth. Don't measure wealth in monetary increments. True wealth is measured by the joy, happiness, and satisfaction in our lives. These give us the greatest of treasures. Robert Collier, one of America's original self-help authors, is quoted as saying, "All riches have their origin in mind. Wealth is in ideas—not money." Don't ignore your dreams and desires; pursue them, and you'll find that inspiration is a naturally occurring element in your life.

2. Think about activities or skills you have learned very easily, naturally, and with enjoyment. Henri Matisse, often regarded as the most important French painter of the 20th century, never painted till his mother gave him a paint box when he was 21, during a convalescence from an operation. During that time, he found his true talent and pursued painting. His artistic career flourished, and rivaled that of Picasso. Only when he had the chance to see his true abilities shine through did Matisse realize what inspired him. Painting came easily to Matisse, once he tried it. What comes naturally for you?

3. What things give you a sense of inner satisfaction, achievement, and fulfillment? What do friends, family, and colleagues compliment you for, or take for granted you'll do well? You should focus in on these things, as they are most likely your true inner strengths. If you do these things well, and enjoy them, inspiration will come naturally.

4. Look back at moments when you felt most alive, happy, and vibrant. What are the things you get completely absorbed in, so much so that you lose track of time? Things that consume your entire focus come naturally, and with enjoyment. As the saying goes, "Time flies when you're having fun." And if you're having fun, inspiration is already there.

---

1. From Jane Firbank's article "How to Find Your Strengths". Please see the Recommended Reading and Acknowledgments sections for more helpful information from Jane and other sources discussing strength finding.

I encourage you to dwell on these thoughts, and delve deeper into this area. Remember why you chose to pursue your dreams, and be sure your dreams involve the desire to create and design before trying to build an inspirational spirit for your design-oriented career.

If you are finding what you do uninspiring, then you will have difficulty applying the other three keys of Innovation, Exploration, and Creation. It's important that you find your inspiration, build your inspiration, and foster an environment where you remain inspired. If not, the symbiotic dependencies between the elements of the *Four Keys to Successful Design* will break down. I think that Johann Gottfried Von Herder summed this up well when he said:

> *Without inspiration the best powers of the mind remain dormant*

Do not let your inspiration go to waste. It's far too valuable, and the key to unlocking so much more of your potential.

You may find that you need to build your inspiration. If so, don't be disheartened. That's natural. The mighty oak was once an acorn that needed the right nourishment from water, sunshine, and soil nutrients to grow, yet the mighty oak did not grow into a giant overnight. Growth takes time, and can be slow. Nevertheless, it is progress just the same. So too is inspiration. If our inspiration is low, and weak, it may take time to build it up and see it flourish. Never be discouraged by this and remember that it is never too late to spark the growth of inspiration.

Stay inspired and seek out that which inspires you. Great things will come as a result.

Before pursuing the next leg of our journey through the four keys, I would like to cap off this first of the four keys with some dessert; some tasty tidbits spoken by some amazing people:

> *The future belongs to those who believe in the beauty of their dreams.*—*Eleanor Roosevelt*
>
> *What man's mind can create, man's character can control.*—*Thomas Edison*
>
> *There is one thing stronger than all the armies in the world, and that is an idea whose time has come.*—*Victor Hugo*

Key #2: Innovation

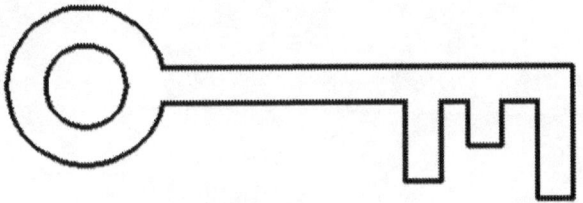

5

An Idea Is Born

o o o o o o o o o o o o o o o o o o o o o o o o o o o o o o o o o o

Imagination is more important than knowledge.

—Albert Einstein

Something sparked your thought; you had inspiration. This notion grew, and became innovation. It could be a brand new gadget never seen before. On the other hand, it could be a better way of doing something. Innovation could be the idea for wingless airplanes, or for lighter ones. Innovation could be the concept of a car that runs on water, or a more efficient one. Innovation could be a new line of children's clothing, or clothes that are just more durable. And innovation could be as simple as finding a way to knock a few days off your project schedule.

Innovation comes in many forms, and is driven by inspiration. The human race, from its beginning, has continually challenged itself to find new and better ways of doing things. If it were not for innovation, we might still be living in caves, wearing animal skins, hunting and gathering our food, and attempting to master the mystery of fire. However, humankind has come a long way, due in large part to innovation.

Nothing seems to impede the forward momentum of progress, and progress takes innovation. Ancient man[1] had the innovation to turn animal bones into tools. The first tool was something completely new to humankind. But then the innovation of the human race went further to use stones, beating and chipping them into tools for cutting, plowing, and hunting. Further innovations inspired by tool

---

1. Reference to "man" refers to humankind, not just the male species. Women also played a vital role in innovations and still do today.

manufacture led to bronze, steel, and other metals to improve upon the initial concepts of utensil manufacture, weapon building, and food production.

The wheels of progress, fueled by the innovation of humankind, have gained tremendous momentum. Like a fast moving train, it seems that nothing can stand in its way. As long as we continue to think and continue to wonder, asking ourselves questions like "Why?", and "Is there a better way?", innovation will continue to accelerate forward.

However, not everyone believes that the forward progression of humankind can continue. Some believe that innovation is nearing its end. After all, we've come so far. Can we continue to advance, and keep the wheels of innovation turning? In 1899 Charles Duell, the Commissioner of the U.S. Office of Patents, was quoted as saying, "Everything that can be invented has been invented." Yeah, right. I don't believe that for a second. How could this "expert" on innovations think that, in 1899, there was nothing left to invent?

Duell's quote always puzzled me. How could anyone think we were nearing the end of the line? I've given it a lot of thought, and wondered what drove this man, an expert in the field of innovation, to say such a thing. Was Duell so shocked by the vast number of innovative ideas that poured into his office that he became overwhelmed and crass? My diagnosis, as humble and opinionated as it may be, is shortsightedness from the lack of inspiration brought on by low employee satisfaction. Duell's inspiration faded away. Somewhere along the road, perhaps due to the overwhelming number of patent applications that flooded his desk, he became uninspired. We may never know exactly what drove Duell to say what he did. In any case, he was wrong...very wrong.

Think of all the things made by innovative people since 1899, following Duell's infamous repudiation. It's astounding to say the least. Think of the automobiles we have today, television, mobile phones, computers, the Internet, breakthroughs in medicine, and space travel. How about the invention of Willis Carrier's first air conditioner in 1902, the Wright brothers airplane in 1903, the Model T in 1908, the cotton picking machine in 1936, nylon in 1938, magnetic recording tape in 1928, or the Polaroid camera in 1947? Think of medicines that were discovered and patented, like the Sulfa Drugs in 1908, insulin in 1922, Penicillin in 1928, and the cardiac pacemaker in 1932.

If Duell was right, our lives would be drastically different than they are today. So why should we today, in the 21st century, ever think like Charles Duell? We take many innovations for granted in our everyday lives. It may appear as though all the "good" ideas are taken, and Charles Duell was right. However, our lives today serve as evidence that Charles Duell was incorrect in his statement.

Innovation, as mentioned in Chapter 1, is a link in the food chain comprising the four keys. Feeding off inspiration, innovation could either flourish or die off, as it did with what appears to have been a disgruntled government employee in the U.S. Office of Patents in 1899. If we do not stay inspired, then our innovation will weaken. If our innovation weakens, so does the quality of our end goal. Innovation has the momentum of humankind's historic achievements. Nevertheless, we need to keep it fueled so that the spirit of innovation is never extinguished.

Innovation is a matter of not being satisfied with the way things are, and always searching for a better way. It's a matter of asking "why?" "Why is this thing built like that?" "Why can't we use X instead of Z?" And of course, "Why not?"

Innovation is a process of questions, sometimes doubt, but always curiosity. Innovation is a line of sight, looking at things from a different angle, perhaps through someone else's eyes. Innovation is the birth of ideas spawned from our inspiration to continually wonder. Once we stop wondering, we become like Charles Duell; our inspiration dries up, blows away on the winds of change, away from our grasp, and forever lost amongst the sands of time.

At times, innovation can seem like plain old dumb luck. Sometimes accidents have led to the discoveries of the most innovative products of our time. Nevertheless, just because we may stumble across a new idea does not mean we just got "lucky." It is true that a blind squirrel can find a nut from time to time. However, it takes more than luck to drive that innovation into something more than just simple thought. You first must be inspired and in constant wonder, questioning everything. Then, when the idea strikes, be ready to grab it by the horns and ride this bull for all its worth. It may still be a long journey to get this idea through the other keys of Exploration and Creation, and you have to be willing to stick it out through thick and thin.

Car tires have an interesting story of innovation beyond mere accidental discovery, via the inspiration and motivation behind a man, Charles Goodyear, who constantly looked for a better way. Goodyear received a patent in 1844 for a process known as vulcanization, which he discovered by accident. This process led to the more durable tires we have today.

Natural or India rubber, as it was known in the 1800s, was of limited usefulness to industry. Rubber products melted in hot weather, froze and cracked in cold weather, and adhered to virtually everything. That all changed one day in the mid-19th century when Charles Goodyear accidentally dropped some rubber mixed with sulfur on a hot stove. He could have become frustrated and merely scraped this spilled mixture of sulfur and rubber off the stove. Having a constant sense of innovation, however, Charles Goodyear saw something different. After cooling the rubber for a short period, Charles Goodyear found that this new discovery was very elastic, despite the fact that it was still quite warm. Goodyear found a resilient rubber, not a frustrating spill that needed cleaning.

Charles Goodyear didn't quite know what happened, but he knew he was on to something; something innovative that would change rubber production for years to come. Later on, scientists unraveled the mystery of this resilient mixture of sulfur and rubber, and built upon this innovation over time. As it turns out, after heating this mixture, molecules in the rubber became cross-linked to one another. Via this "cross-linking" the molecular structures become loosely bound to each other. The sulfur links kept the long molecules from slipping by each other at high temperatures. This keeps the rubber resilient so that it returns to normal after stretching, much like a simple rubber band or elastic. And best of all, it wasn't sticky. This new process of vulcanization connected the strands through the sulfur links so that the interconnected molecules retain their orientation when exposed to high temperatures, like the friction of automobile tires on the road.

Granted his first patent for this process in 1844 (patent #3,633), Charles Goodyear fought numerous legal battles over patent infringement, and later died deep in debt in 1860. Nevertheless, his innovative idea, spawned by an inspirational thought, led to the improvement of the car tires that we know today.

The key to innovation is not just being in the right place at the right time. It's a constant process of curiosity, and consistently looking for better or new ways of doing things. You need to have inspiration so that your eyes can remain open, and aware. It is when you are truly inspired that you will be genuinely innovative. Goodyear could have just scraped off the spilled rubber from the stove, but his inspiration just wouldn't let him do that. He took things one step further and an idea was born; an innovation that made a mark in transportation history.

Be inspired, be constantly curious, and forever wonder. Do not force innovation. Instead, by building your inspiration, being persistently inquisitive, and not looking at mishaps as failures, you will ensure that innovations will often find you.

Now, let's roll up our sleeves and get down to the nitty-gritty of this second key: Innovation.

6

Necessity is a Mother

Innovation is often born from necessity, the famous mother of invention. Progressing through time, humankind has come to realize that the things we want or need are not far from our reach. Prehistoric humans may have wished they had personal transportation, but of course, the automobile was millennia away. Nevertheless, as time moved on, the desire for such things as personal transportation evolved from necessity, which gave birth to saddled horses, bicycles, and of course, much later, the gasoline-powered car. Innovation is birthed from necessity bred with inspiration. As time moves on, subsequent births evolve based on past experience and the availability of past innovations. Such is the case of the auto, going from the original barefoot man, to animal-assisted transport, then to mechanically assisted transport.

The Internet has served as a major turning point in the evolution of humankind's knowledge, expediting information into our homes and offices so that we can progress. When I was in high school, I could only have wished for not only my local library to be as close as my living room, but also libraries from around the world. In more recent years, the Internet has served as a catalyst in the birth of many new ideas based on necessity. We can browse bookstores online, we can do our holiday shopping online, and we can communicate and exchange information at the click of a mouse.

Besides just surfing the Internet, I love to do the real deal; jump into the coastal waters, paddle my longboard out past the breakers, and ride the waves. I've lived in California since the mid 1980s, and surfing was one of the first things I learned to do when I came to the Golden State. Moving from Ohio, and already in my early 20s at the time, it wasn't exactly a smooth progression for me to paddle out into the ocean, balance myself on a surfboard, catch waves, and not fall down. Needless to say, I got hurt quite a few times. I busted my nose once; it took a few stitches and a bit of surgery to get my snout straight again. In addition, I've had my share of bumps and bruises from other wipeouts, mostly from taking on conditions beyond my limits. I'm not an expert surfer today by any means, but I thoroughly enjoy the sport. I have learned my limits and am in much better shape for the adventure. I also learned some key elements surrounding the science of the sport that led me to an inspired innovation.

One of many surprises during my initiation into surfing was what appeared to be the unpredictability of the ocean. One day I would go out, and the waves were weak, perhaps one to two feet. Then by the next day, the surf could be pounding the coast with six-to-eight foot waves, and at times, even bigger ones. The area in California I like to surf occasionally sees 15 foot waves (which are way out of my league) and during big winters sometimes 20 feet, although that is rare. Still, I know my limits and I don't want to take the time to drive to the beach if I'll be facing conditions not suited for my skill set (and sense of reason).

Tide times were somewhat of a mystery as well, as were wind and weather patterns that varied from my inland residence to the coast some 30 miles away. These factors play a vital role in surfing and, like wave size, are crucial pieces of knowledge I like to know well in advance.

To tackle these unfamiliar ocean elements, I went in search of answers. At the time, during the early to mid 1980s, there was little information available to the public on surfing conditions and surf forecasts. The local paper would have a small snippet, mainly geared toward boaters, that would give the tide information and a very rough estimate on swell height for a given day. Later, I found a pay surf forecasting service that I could call on the phone, and for a per-minute price, I could listen to a recording of the current conditions. This service also had something more important to me: a forecast of what was to come in the way of surf over the next few days, or possibly longer.

Knowing this information beforehand was essential. Not only would I save myself valuable time, but I had a better heads-up on safety issues as well. What I also found extremely useful was the ability to plan my week. Well in advance, I could arrange for a day off from work when I knew some good surf was heading to the coast. However, my daily phone calls to the pay-per-minute phone service came at a price, a high price. Additionally, the information I was paying for was not that accurate.

After a few years of doling out dollars for the pay-by-phone surf forecast information, the Internet started to make its appearance in most American homes, including mine. It wasn't long before I realized I could access weather maps, swell models, wave analysis models, buoy data, and a variety of free information, which I could use to track storms across the Pacific. Using this publicly available information from sources such as NOAA and FNMOC, I was able to calculate the arrival of the swells[1] that would hit California beaches, and estimate their sizes as well. This was a lot cheaper than calling a 900 number for the information, and I was finding my forecasts to be far more accurate as well. Soon after, I stopped calling the pay service for my surf forecasts and began making them myself, free of charge.

Before long, I was hooked on the "ham radio hobby" of the late 20th century: the Internet and all its cool gadgetry. I was turning into a bona fide computer geek. I started a web site dedicated to—you guessed it—surfing. I had an open community environment and exchanged e-mails regularly with a variety of fellow surfers from around the world. Given the Southern California theme of my web site, the majority of visitors were from the California regions, ranging from San Diego northward through Santa Barbara.

As the number of contacts grew, I started a free e-mail newsletter to keep all my web site visitors up-to-date on the latest and greatest surf-related stuff I was finding on the Internet. Occasionally I would mention news regarding the more significant swells I was tracking on their way to the California coast; when they would hit, how big they might be, and other various conditions to anticipate.

Technology had me inspired; it always has. With the hobby of the Internet, and my love for surfing, combined with camaraderie from the audience visiting my

---

1.   In the surfing, boating and many other maritime communities the term "swell" refers to a day or more of surf from typically a storm driven system. When a "swell" hits the coast, there are plenty of waves for a day or more.

site, my inspirational appetite was being gorged. I had some innovative ideas for the web site, but none as strong as a suggestion that I received one day in an e-mail from one of the web site visitors: the information regarding a particular swell I predicted was great, and he would be willing to pay me for that kind of information.

The lights went on in my head and the wheels of inspiration were turning. It was the birth of my surf forecasting service: WaveCast® at wavecast.com

Originally, I provided this service seven days a week. A member could join for a small price using a credit card online, 24 hours a day. I wrote software to automate much of the site administration, and I created the forecasts manually each day. There was a lot of work involved, but I loved every minute of it and was making some decent cash at the same time.

After WaveCast® was taking in members for a few months, a local paper, the Star, contacted me for a story. They came out, interviewed me, took some pictures, and I got my 15 minutes of fame. Then competitive newspapers like the Los Angeles Daily News wanted a piece of this new literal "Web Surfing" action, and came out to interview me for their own articles. Even the Associated Press took on a story of this new, one of a kind online business. I was getting a little more than my 15 minutes (but not much more).

During each interview, I was asked a similar question, "Was that my real name (Cool), or did I make that up?" The question raised some doubts due to the nature of WaveCast®, a surfing site, being a *cool* thing. Truth of the matter is, I *was* born with that name, so was my father, his father, and our relatives in the Old World. I was more than aware of the parallelism between my name and the theme of WaveCast® but I did not want to use my name as the driving force behind a good idea. WaveCast was to be successful not by catchy phrases, but by the content that it provided. The content was king. This is what the paying members wanted. No one would pay $6.00 a month to belong to some site just because of the owner's last name. My mother is a great woman, and although we have the same last name, <u>necessity</u> was the true mother of WaveCast®, and the innovative idea for this new kind of service on the Internet.

Today, working with wetsand.com, the WaveCast® service is able to provide a more innovative resource. Using an online store for revenue generation, the new business model allows visitors to access the site entirely free; an innovative idea that some competitors deem provocative. Through constant improvement, we've automated more of the data for readers to gather surf and wind information from many surf breaks around the globe. It all started with an inspiration: my love for surfing and the Internet. Innovative ideas were born to bring this service, Wave-Cast®, to the masses, using a variety of web technology while exploring the various options that lead to the creation of WaveCast®, and its successful partnership with WetSand.com

Necessity is truly the mother of innovation. Even today, I want to know if the drive to the beach will be worth my time. Will the surf be too big or too small? Will the winds cooperate? What is the best time of day to go out for a wet session? I *need* to know these things, just as many other surfers who have busy schedules want to plan not only today, but also perhaps days or weeks ahead. My necessity to know information, fed by an inspirational spirit, mothered an innovative idea: WaveCast.com, which after careful exploration, was created, constantly improved over time, and remains successful to this day.

Once again, we can see how the four keys, Inspiration, Innovation, Exploration, and Creation are bound to each other in a tightly woven fabric. Removing a single thread could unravel the entire structure. The four keys are truly dependant upon one another. If it were not for inspiration, would I have had the idea for a web site? If it were not for innovation, and lots of exploration, nothing would have come of the idea, and nothing would have been created. Furthermore, if I had lost my inspiration at any step along the way, the whole idea could have collapsed. But it didn't. The four keys maintained my goals, and led to a successful design.

I'm reminded of a famous quote when I think of necessity and innovation. You may have heard this one before, the "the six words to success":

Find a need and fill it.

There are numerous stories where these six words to success made a foothold. WaveCast® was just one of many. As you foster your innovation, keep these words boiling on the back burner of your mind, and remember a famous quote from 1963:

We need men who can dream of things that never were.—John F. Kennedy

The world truly needs men and women who can innovate, and bring to fruition the things that never were. Be aware of necessity and the needs of humankind, feed your inspiration and be constantly curious. Innovation will naturally flow as a result.

You may be thinking to yourself that this is all a best-case scenario; that it just sounds too easy. What happens when opposition stands in our way? How can we stay inspired and innovative when discouragement looks us straight in the eyes? Not to worry. Nothing can stand in the way of truly inspired innovative progress. Let's take a closer look as we explore the next chapter in the key of Innovation.

7

Expansion from Discouragement

Change your thoughts and you change your world.

—*Norman Vincent Peale*

Naysayers are inspirational-appetite suppressants. They are a resistive force that, if left unchecked, will impede forward progression. Although a pessimist may have a good argument regarding a new idea, innovative thought requires a filtering mechanism to weed out discouragement, yet encourage constructive feedback. Innovation grows from encouragement and, if approached from the right perspective, can expand from it as well.

Throughout time, inspirational defeatists have said some amazing things. Lord Kelvin, president of the Royal Society of England in 1895 was quoted as saying, "Heavier-than-air flying machines are impossible." Tell that to the Wright brothers, who only 18 years later made their historic 120-foot, 12-second flight at Kitty Hawk.

Once old Lord Kelvin was proven wrong, there were more "Doubting Thomas's" to pose as obstacles in the development of modern aircraft. Marechal Ferdinand Foch, Professor of Strategy at the "Ecole Superieure" de Guerre military school in Paris, was noted as saying "Airplanes are interesting toys but of no military value." That one still puzzles me. I don't know who had a lower inspirational appetite, Charles Duell from the Office of Patents in 1899, or this scoffer Foch.

The list of discouragement goes on. Think about the computers we use today. Think about how many are in the workplace, as well as the home, and what an integral part of society computers are. Then think about what Thomas Watson, the chairperson for IBM in 1943 said when referring to computers: "I think there is a world market for maybe five computers." Five? Only *five*? Obviously "big blue" has had better visionaries since then, and since 1943, it has flourished in the computer market. But at the time, 1943, I'm sure that IBM's top brass were not inspired by computers, for the company's primary market was business machines. Computers didn't play a large role in 1943. As time went on, that inspiration changed.

No matter what others may say in a negative connotation towards your innovations, discouragement should not be something that impedes your forward momentum. Instead, we should be like sponges and soak up all the discouragement we can. And like sponges soaking up water, we need to expand and broaden our innovative thought.

Discouragement has two primary components. There is a positive component; someone else's viewpoint. Then there is the negative component, which may contain pessimism, cynicism, sarcasm, skepticism, or all of the above. With practice, you can learn to separate the two components. Like peeling an orange, you will be able to differentiate the sweet fruit inside, yet throw away the bitter rind. Learning to deal with these two portions of discouragement can be disheartening. But they are not insurmountable.

I like what Eleanor Roosevelt once said:

> *Do what you feel in your heart to be right—for you'll be criticized anyway. You'll be damned if you do, and damned if you don't.*

Eleanor was onto something there. Criticism is like a naturally occurring phenomenon. It's intrinsic to the world of innovative thinkers, and cannot be avoided.

So let's face it. We're going to be criticized, and we're going to get our fair share of discouragement along the way. Not everyone is going to be as excited about your ideas as you are. Your innovations, like newly born fawns, have vulnerabilities. Each idea you have has a hard solid back, strengthened by your inspirationally fed thought. You have a great idea and you stand behind it. You feel it is sound and has merit. However, your innovations also have a soft underbelly that can put them at risk, especially when they are young.

Your innovations will be most susceptible to discouragement during their infancy. Anything new is often greeted with skepticism, doubt, and at times an unwillingness to change. As history tells us, people are not always open to change. Roger Bacon, recognized as one of the earliest advocates of modern science, was sent to prison in 1278 for his "radical novelties" of teaching. Many people during the 11th century, including the Franciscan order who imprisoned Bacon, felt that Bacon's ideas were heresy, and that he posed a threat to society at the time, all because of his innovative way of thinking.

Although you are less likely today to be thrown in the slammer for your innovative thought, people today are still not always ready for change. One such instance in modern times, in the mid 1970s, is when Steve Jobs and Steve Wozniak went out on their own to make the first Apple® computer. At the time, many saw this as a revolutionary innovation: a personal computer for use in the home! With little money to invest in this project, Jobs and Wozniak tried to sell their new innovative idea to companies such as Atari, HP, and MOS Tech. However, these larger companies were apprehensive at the time, and not open to the change proposed from these young upstarts.

Steve Jobs, in his own words, is quoted describing these discouraging attempts:

> "So we went to Atari and said, 'Hey, we've got this amazing thing, even built with some of your parts, and what do you think about funding us? Or we'll give it to you. We just want to do it. Pay our salary, we'll come work for you.' And they said, 'No.' So then we went to Hewlett-Packard, and they said, 'Hey, we don't need you. You haven't got through college yet.'"

Well, the rest, as the say, is history. Apple® computer's success, initially attributed to Jobs and Wozniak, is a demonstration of how innovations, during their vulnerable infancy, are susceptible to discouragement. Nevertheless, as history tells us, discouragement does not mean defeat.

So, how do we deal with this discouragement? And how do we soak it all up and expand from it? The best way I have found to deal with discouragement is to attack it with what I like to call "discouragement-management."

To understand discouragement-management, it's good to be familiar with a fundamental of human nature: cognitive distortion. Cognitive distortions are a way your mind interprets sensory input in a negative form and not necessarily its intended form. Someone says or does something that provides your mind with input. How your mind processes this input will dictate how you feel and ultimately react. If someone throws a batch of discouragement your way, you may naturally take this as defeat, hang your head low, and give up on your idea. However, doing so would be a form of cognitive distortion, which you can control to your advantage. Discouragement need not lead us to despair; instead, it can be used as a building block on which to construct positive improvement in our innovations. Tackling these cognitive distortions is the first step in understanding how to tame the beast of defeat and manipulate it to your advantage.

When discouraging events occur, perhaps some negative comments from your peers, your mind interprets these events amid a series of thoughts, known as your internal dialogue. This internal dialogue continually flows through your mind. In turn, this internal dialogue creates your feelings and moods. However, the internal dialogue is the intermediary component in this process.

There are three elements involved: the discouraging events, your internal dialogue (your thoughts), and then your feelings and moods, from which you react. The events, however, did not create your feelings and mood. Instead, your internal dialogue, the middle man in this process, gave you a perception of what the events mean. You cannot control the events, but you can control the internal dialogue, thereby controlling your perception, the way you feel about it, and your ultimate reaction.

There are ten common cognitive distortions recognized by most analysts[1]. I've found three of these distortions that relate very closely to discouragement, and have found ways to deal with them as well. Let's take a closer look at these distortions:

- **All or nothing thinking:** This is where everything is either good, or everything is bad. You can't see any grey area. If one review of your design doesn't go your way, you tend to think that you are a failure, and that you never have any good ideas.

- **Overgeneralization:** This is where a single negative event spawns a feeling of constant defeat. You arbitrarily conclude that one thing that happened to you will occur repeatedly. If your prototype of a new circuit blows up, you think that everything that you ever try again will do the same.

---

1. Please see the Recommended Reading section for sources that describe in more detail all ten cognitive distortions.

- **Mental Filtering:** This is where you pick out a single negative detail in a situation, dwell on it exclusively; thus perceiving that the whole situation is negative.

All of these cognitive distortions are somewhat related, and you may be able to relate these to some discouragement you've encountered at some point in your life.

Let's take a look at the first distortion, "All or nothing thinking." Say you're working on an idea for a new way to sort mail at the Post Office. You think this is a fantastic idea that would save time and money; something the Postmaster would appreciate. You present the idea to your local Postmaster and he doesn't really see your idea as useful. With the "All or nothing thinking" cognitive distortion, you might feel that you are a failure, that everything you do is crap. You may think to yourself, "Why can't I ever do anything right? No one ever likes my ideas." Now step back for a moment. Is this true? I'm sure you do some things right, so you can't say that you *never* do. Moreover, how can you say that no one likes your ideas? You must have had some good idea at some point in time.

Life is rarely completely one way or the other. We don't live in a binary world where things are either all good or all bad. Between the colors of black and white are numerous shades of gray. Have you ever had anything "perfect"? Even the greatest of things tend to have flaws. A "perfect" meal might not be served at *exactly* the right temperature. A "perfect" body still has a wrinkle or two. I'm sure this book has flaws in spelling and grammar. I'm sure it's far from perfect. And what is a "perfect" design? I have yet to see anything made that I can't find fault with (even my own book). The fact is that nothing in this world is *exactly* perfect. Instead, this world is composed of *nearly* perfect things.

Let's revisit the example confrontation with the Postmaster. This person may have been crass and very abrupt in his approach with you. That could be very discouraging, and you might have the natural reaction to take this as a defeat and throw away your idea completely. That would probably be a big mistake. Look closely at this discouraging event again. The Postmaster may not be very adaptable to change. He may also have been very busy at the time. In addition, what exactly didn't he (or she) like about your idea? Was it the whole idea that he/she didn't like, or just a portion of it? I'd lay dollars to donuts that at least some of the idea was worthwhile. If the Post Office won't buy into your design, there may be someone else who will. Perhaps other businesses could utilize some derivation of your idea based on the good portions that the Postmaster couldn't criticize. Sometimes we need to listen to what others "don't" say.

Your internal dialogue, in processing the discouraging exchange with the Postmaster, could lead you to a terrible feeling about yourself and your design. On the other hand, you could process the events to perceive the positive virtues of those events. This would inevitably lead to higher a higher level of inspiration. Being further inspired, and using the positive input constructively, you could expand your innovative thought and quite possibly design something better.

The second cognitive distortion previously mentioned, overgeneralization, makes us fall into a similar trap as the "All or nothing thinking" distortion. In this case, when you get the rejection of an idea, you see yourself in a never-ending pattern of defeat. You might think such things as "When it rains it pours." Or you might think, "They don't like my idea now, and they never will." But are your assumptions true? I highly doubt it. Just because something happens once doesn't necessarily mean it will continue to occur. Everyone has good ideas, great ideas, and sometimes bad ideas. Nevertheless, I have yet to meet anyone who has nothing but bad ideas.

The last cognitive distortion I wanted to discuss is Mental Filtering. Similar to overgeneralization, this distorted way of perceiving discouragement places you in a spiral of defeat. Take the example of the Post Office idea we discussed regarding the "All or Northing" distortion. Say the Postmaster said that it was a great idea, but they could never use such a thing. If you are Mental Filtering, you might take the single connotation that "they could never use such a thing" as meaning that your idea is garbage. However, didn't the Postmaster also say, "It was a great idea"? In this case, the idea probably is good and someone else would probably buy into it. Nevertheless, if you allow the Mental Filtering to take hold, you will face the discouraging event with defeat due to one portion of the event. But this shouldn't stop you from going forward. The idea may need to change course, but it shouldn't be thrown away.

Understanding cognitive distortion will help in managing discouragement. By being aware of these distorted thought patterns, you will be able to differentiate between the two components that comprise discouragement. There is always a positive component and a negative component. However, the two are not always equal in weight. The key to using discouragement management is to step back, be cognitive of the entire situation, then divvy up the positive and negative components.

Let's look at the Postmaster example again and see how we can utilize discouragement management to deal with the situation. If the Postmaster had said, "I like this idea, but I don't think we have a use for it here," then you could judge the positive portion as perhaps 90 percent, and the negative component as only 10 percent. You realize that you have a very good idea (the positive 90 percent), but you were showing it to the wrong audience (the negative 10 percent). In this case, all you need to do is look for another audience.

But what about the other side of the coin? Let's say the Postmaster's response was something along the lines of, "This thing is completely useless. I wouldn't waste my time with this piece of junk." In this case the positive component is more in the 10 percent range, and the negative in the 90 percent range. Listen closely, however, and perhaps consider showing someone else your innovation. Such a negative kind of response could mean you're dealing with an isolated incident. Remember that having one negative encounter doesn't mean it's over.

These two examples lean toward the extremes of what may have happened when you proposed your innovation to our make-believe Postmaster. We need to look at that bigger portion of the pie; the more common and highly populated grey area between the extremes. Let's say the Postmaster's response was, "I don't know. That seems nifty, but it won't easily integrate into our system. Thanks for coming by, but we're not really interested." This response has mixed signals. Can you see where the cognitive distortion is in this example? One cognitively distorted perception could be defeat because the Postmaster said he wasn't interested. But look closely at what else he said. The first thing he said was that it was "nifty," and then he said it wouldn't integrate well. Judging by this, I'd conclude a positive component of 75 percent, and only a 25 percent negative component. Your idea is good. He called it nifty. He also didn't say it was crapola. He did say, however, that it would not integrate into their system. If you tweak your design a little bit to better fit the Postmaster's integration objectives, you can get back some of the 25 percent that fell into negative territory.

Learn to utilize the positive aspects of discouragement, no matter how small they may be. Remember that you are in control of the sensory input you receive. You can make a choice in the processing of the input, and the resulting outcome. Learn to separate the fruit from the rind of discouragement, and maintain focus on your goal. Use the opportunities that discouragement throws your way by applying the positive aspects of discouragement to your advantage.

Learn from all the input others have to offer and don't feel disheartened by it. Keep a positive attitude and maintain your inspiration. Remember that others will likely criticize your ideas, but there are those willing to accept them as well. Utilize discouragement as a means to enhance your innovations, constantly expanding your horizons and inspirationally induced innovative thought.

8

Avoiding Next Bench

o o o o o o o o o o o o o o o o o o o o o o o o o o o o o o o o o o

Experience teaches only the teachable.

—*Aldous Huxley*

Throughout the engineering community, there is a term known as "next bench marketing," adopted by a number of companies during the Neolithic period of electronics, also known as the 1970s. During those bellbottom days, next bench was a process of one engineer taking a newly minted idea for a product or feature he or she came up with while working in their dark corner of the cube farm, and sharing it with his or her closest peers. If you came up with an idea, you'd pop up from your chair, hang your arms over the cube wall, and ask the person next to you if your idea had merit. If that person occupying the cube next to yours answers yes, then you obviously must have a winner. If you get a negative response, then you probably didn't explain it well enough, so you try the person on the other side of your cube. Next-bench may have had some merit for oscilloscopes, voltmeters, and similar items used by engineers and other geeks. However, next-bench doesn't work well when we are innovating for the masses outside our cube sector.

There is no doubt that it's a good idea to get the opinion of other colleagues when you're working on a problem. However, I've seen this develop a vacuum from which the engineers cannot escape. Getting only the opinions of your immediate peers can be myopic. You will tend to get the input of only one facet of the design, input that, coming from an immediate peer, will more than likely be very similar to your perspective. This is not to say that next-bench is a completely flawed concept. It has a purpose, but needs to be taken in the right context.

Let's face it. If we left the entire design of a product to us cube-dwelling day laborers, we would probably end up with something that reeked of Star Trek and behaved like DOOM 3. This is not to say that us engineers, designers, architects, and inventors never have ideas that warrant merit. However, it is important to realize that without the stimulus of others outside our immediate circle, we will miss some valuable information.

Communal development throughout world history has proven that societies with higher populations established the most successful sociological infrastructures. These societies dealt in trade with other neighboring societies, and exchanged ideas. The evolution of Polynesian societies is an excellent example of this, and shows us why next-bench thinking could hinder our innovations.

A study in the migration of various Polynesian societies throughout various islands in the Pacific during a span of time from approximately 1200 B.C. to 500 A.D. shows that the most secluded islands, like the Chatham's, with lower populations, had little interaction outside of their microcosms and developed slower as a result[1]. The less populated and secluded societies were not as successful in their developmental progression of agriculture, tool development, or economic growth. Lower populous regions had more simplistic tools, rudimentary agriculture, and elementary housing. These more secluded, less populated islands suffered from the lack of combined thought and ideas.

Higher populated islands, on the other hand, like Samoa, Tonga, and Hawaii, flourished. These islands with higher populations had more interaction amongst themselves, as well as other neighboring Polynesian societies. These higher populated societies advanced more quickly as a result of venturing outside their immediate circle, and welcoming the involvement of others. It was collective thought from a versatile range of individuals, not just a small isolated microcosm.

In short, we learn from each other and we need the input of others. However, if we stay only within our secluded islands, and never venture outside to acquire the involvement of others, we'll inevitably lose valuable contributions that could improve upon our innovations.

A common exercise that I've used on a number of software projects is peer review. This is where, on a regular basis, our development team will join engineers from

---

1.   Referencing context from "Guns, Germs, and Steel, The Fate of Human Societies by Jared Diamond"

outside groups to talk about the software we're currently developing. We will look at the code for various things. We'll analyze such things as algorithms, design patterns, best practices, and of course syntax. This is a difficult thing to do sometimes if it's your code under the microscope. After all, that code is reflecting your innovation. To have that come under scrutiny can be very humbling. But what is the alternative? To develop your software in a vacuum? To think that your ideas are the only ideas? When you think about it, the alternatives sound conceited.

It's true that you were inspired and developed something innovative, no doubt about it. It was your idea, and you deserve credit for it. Nevertheless, your peers, as well as those outside the group have inspiration and innovation as well. Moreover, each of us has a different inspirational appetite that can lead us into various paths of innovative development; thus broadening our horizons.

Something that also helps in peer reviews is to have someone wear a black hat. This person, who should be different for each review or meeting, wears the black hat to designate himself or herself as the "Devil's Advocate." This person, no matter what, is a designated faultfinder. This person is to question everything and anything. Since you know that this person's role is to do just that, it's not a surprise when you hear negative input. It also allows the Devil's Advocate to think outside the box, purposefully looking at things from a different angle. By performing this black-hatter exercise in your peer reviews, you gain the insight needed for worst case scenarios. By having a different person wear the hat at each review, each team member is forced into a different train of thought.

The four keys presented in this book have a careful balance. So does the input and influence of others. Although the majority of input tends to be constructive, bear in mind that at times you may face some malicious attacks. We all can admit to having a bad day. Perhaps we didn't sleep well the night before, got up late, had a flat tire on the way to the office, or were stuck in traffic. It's on those bad days when people may not give their best advice. Also, there are those rare times when negative input is unjustifiable due to a person's jealousy, anger, or permanent bad attitude.

Recall from the previous chapter, Expansion from Discouragement, that when you're faced with discouragement, you need to analyze the input and consider the positive and negative aspects of it. Extract the positive portion of the input and allow it to reinforce your innovative thought. Perhaps your innovation needs a bit of tweaking in some areas. That doesn't mean it's garbage. It's still innovative, but perhaps could use a bit of polishing. Remember that one man alone did not design the space shuttle. It was the inspiration and innovative efforts of various teams that made it happen.

Always remember the fifth habit cited by Stephen Covey in his book "The 7 Habits of Highly Effective People":

Seek first to understand, then be understood.

Once you understand the constructive aspects of the input, you can better communicate with those giving the input.

Getting beyond next bench is more than just gathering input. It's crucial to know all aspects of the requirements your innovation will impose. All development processes have (or need) a requirements gathering phase. To gather information from the customer, larger companies (and even some smaller ones) will assign a marketing representative to take the lead in this role. This representative will compile the requirements a product must meet if it's to be a successful design. Marketing then delivers the requirements to the engineers, who in many cases then take over and do nothing more than develop the product. It's true that without input from the market, a product, house, building, or any other design may miss its mark. Marketing input is valuable. Nevertheless, as an engineer you too have requirements to gather. These may simply be getting more details from marketing. It could also be a matter of asking questions not yet addressed. Enhancing the requirement gathering process will broaden your understanding, and help to deliver a more creative and innovative design.

Before starting a project, and many times during one, encourage the feedback of others. When you first have an innovative idea, toss it around to your peers as well as some others outside your circle. Perhaps you could run it by a customer or two. As time goes on, take a step back and get additional feedback.

One of the worst things I have seen in next-bench-only designs is the lack of outside input. I recall an application programming interface (API) that one group was developing that dragged out for over a year before getting canned (along with some of the engineers working on it). The idea they had was to improve the speed of this API, and its interface to the equipment it would control. The group developing this innovation was very tight knit. They all worked together very closely, and stayed well within the confines of their own sector of the cube farm. Rarely did they ever venture outside their microcosm to involve other teams.

Management, leery that the project was taking too long, even for the first prototype, made this secluded API group venture off their island and demonstrate this new API design to other engineering groups for feedback. The input was not positive from the other groups, and the API improvement team went back to their camp and stayed there.

About six months later, management once again asked for some progress, and proof that this API design would work. Long story short, it didn't. Management forced the API group to circulate what little they had developed to various engineers for evaluation. It turned out to be a disaster, and even the mention of the project's name became a symbol of something bad. The code was a mishmash of various languages, there were memory leaks, it constantly crashed, the memory footprint was unwieldy, and the maintenance was difficult. The API, when it did run, was slower than what we already had, and cost more to make. Syntax was awful, and no one could quite figure out how to use it.

This API team had gotten very myopic and secluded on their little island hidden in the far reaches of the cube farm. Without getting input from the outside, their next-bench approach led them to failure.

Innovation is not a lonely ride on a desert highway. It's a party with friends who mingle and bring their strength through numbers. The whole is greater than the sum of its parts. Invite the participation of others during your innovative journeys through design, and your ideas will flourish beyond your expectations.

9

Mouse Traps

o o o o o o o o o o o o o o o o o o o o o o o o o o o o o o o o o o

Engineering is the art or science of making practical.

—*Samuel C. Florman*

Being innovative doesn't mean you need to reinvent the proverbial wheel. Quite often, you will find there's a way to make something better. You'll find a way to make a better mouse trap.

Bear in mind that even new innovative ideas are sometimes just new takes on old problems. The invention of the gasoline-powered automobile by Karl Benz of Germany in 1885 was a new idea. However, it was really an idea to make something better: personal transportation. The actual history of the car dates far back to the days of Leonardo Da Vinci, who drafted up theoretical plans for a motor vehicle. Isaac Newton also drafted up such an idea, though neither Da Vinci's idea nor Newton's idea for a car came to fruition. Innovation by Nicolas Cugnot in 1769 lead to the first steam-powered vehicle used by the French Army to haul artillery up high hills at a whopping speed of 2 1/2 miles per hour. Other designs later used steam to drive vehicles, but the steam engine was a bulky and heavy monstrosity.

Karl Benz, of Daimler-Benz[1] fame, was inspired in the late 1800s by the thought of personal transportation, and had the innovation to make a better mousetrap. Around 1883 Benz, living in Mannheim, Germany, started "Benz & Co." His company produced industrial engines, including various two-stroke engines that he invented and patented. He later heard of a man, Gottlieb Daimler, who was working on a four-wheeled vehicle. Daimler inspired Karl with this new

---

1. Daimler-Benz later rolled out cars we know today as Mercedes Benz

approach to personal transportation. So Karl Benz went to work on his own "motor carriage," with a four-stroke engine.

Benz knew he could improve upon the old, clunky steam engine design, and in 1886 received a patent for what is considered the first gasoline-powered motorcar in history. His design was a simple four-stroke, single-cylinder, water-cooled engine mounted to a three-wheel vehicle. His innovation was truly a better mousetrap than the old steam engine.

Benz's gasoline-powered auto inspired other innovators such as Panhard & Levassor in 1889, and Peugeot in 1891, to manufacture their lines of cars. There was also Charles and Frank Duryea of Springfield, Massachusetts, who in 1896 built their version of the auto, the Duryea Motor Wagon.

But of course the most popular innovator in the history of the auto is probably Henry Ford. Ford didn't reinvent the proverbial wheel, but he did make a much better mousetrap. Ford's innovation was the first most productive conveyor belt assembly line used for mass production of cars, spitting out an astonishing 10,000 cars every 24 hours back in 1925.

Today, car manufacturers continue to improve upon the auto. Think about air bags, power steering, cruise control, ABS brakes, power windows, and more. Automakers continually strive to make a better mousetrap.

Even though ideas like the auto can be considered as improving on an existing idea, there are some ideas that are more revisionist; a radically different take on something. Such an innovative idea is behind the story of the fountain pen, and later the ballpoint pen.

The period of 1880 to 1900 saw an abundance of fountain pen inventions from innovators around the world. Many were not practical, but over 400 patents were granted during that 20-year span. The first practical fountain pen is credited to Lewis E. Waterman, a 45-year-old American insurance broker, in 1884. The story goes that Lewis was getting ready to sign a vital contract on a building site and had bought a new fountain pen for the occasion. The contract was on the table, and the pen was in the client's hand. Once, twice, and even a third time, the pen refused to write. Then, adding insult to injury, the defective fountain pen made an ink blot on the important paperwork, ruining the contract. Waterman scrambled back to his office and quickly obtained a fresh contract, then returned to the site. But it was too late. A rival broker had beaten him to it and the client had signed a contract with the competitor.

Waterman learned from that event and, having an innovative mind, went on to design his own fountain pens, manufacturing them in his brother's workshop. His design was unique in that it was based on the physical force of "capillarity," where air replaces the ink used, giving a smooth, even, blot-free flow. Waterman patented his new pen in 1884. He built a better mousetrap, but more improvements on this idea were yet to come.

Lewis Waterman died in 1901 and his son Frank took over the successful pen business. The younger Waterman added a clip to the cap in 1905, which was a nice improvement, but getting ink in the pen was still a messy process. Waterman's director, M. Perrand, who invented the ink cartridge, refined this refilling process. He put the ink into a small glass tube with a cork stopper. The concept was patented in 1935. Waterman's company was still making better mousetraps.

Then Laszlo Josef Biro, a Hungarian living in Argentina, was working on the idea of putting a ball in the tip of a pen to avoid the pen being clogged from fast drying ink. In 1940, Laszlo had to flee from the Nazis, first to Paris then to Argentina in South America, where he and his chemist brother Georg, took out a patent in 1943 and made the first commercial ballpoint pens. The British Government bought the rights to this patent, as these pens were ideal for RAF Aircrew use. Not only were they more rugged, but they also wrote at high altitudes due to their consequently reduced pressure, whereas fountain pens flooded.

Innovation is the introduction of something new. A new idea can be an improvement of an old idea. The trick is not to be discouraged because an idea already exists. Just because man landed on the moon once doesn't mean he won't be back again with better spacecraft. Like the history of the pen, constantly ask, "Is there a better way?"

Not all mousetraps need improvement. Sometimes, as innovative designers we like to think of new ways of doing things, *our* way of doing things. Even though many things could use improvement, don't be afraid to utilize those things created by other innovative developers. A good example of this is the use of third-party components in software design. Let's say that you're designing a new software program. To distribute this program, you want to ensure that you can license it to protect your costs and redistribution rights. You might think of various encryption methods that you could embed in your code that would lock the software to a user's computer. However, there's no need to reinvent the wheel when there are plenty of other software vendors who make licensing tools, which you can buy and implement into your own software.

By making use of an existing component that you don't need to reinvent, like the licensing component example, you gain a variety of benefits. You:

- **Save time on your design**. With the work already done for you, you should be able to just pop this puppy into the works, and get on with it.

- **Maintain content concentration and focus**. The licensing component was merely an element to help protect your innovative idea. If you were to detract from what it is you do best, the overall product quality could be compromised. Stick to what you do well, and let others stick to what they do well.

- **Obtain expertise and professionalism**. The third-party vendor has a specialty in the field you are looking for. Utilize it.

- **Acquire maintenance and support**. Never forget about tomorrow. Once your idea has come to fruition and is in use, you need to support it. Customers will have problems, and everything inevitably will break at some point in time. By using a third-party vendor when you can, you take some of the support burden off yourself.

You can think of the world as a collection of building blocks. Can you use the available blocks, or do you need completely new ones? Innovation is truly a new idea, but it is polished and refined by practicality. Invent a wheel if it doesn't exist, but reflect on the benefits of making a better mousetrap.

To close out the key of innovation, I'd like to share with you some famous quotes by a man who I feel stands out as one of the greatest innovative minds of all time, Thomas Alva Edison:

> *I never perfected an invention that I did not think about in terms of the service it might give others…I find out what the world needs, then I proceed to invent….*

> *Because ideas have to be original only with regard to their adaptation to the problem at hand, I am always extremely interested in novel ideas others have used successfully….*

> *We have merely scratched the surface of the store of knowledge which will come to us. I believe that we are now, a-tremble on the verge of vast discoveries—discov-*

eries so wondrously important that they will upset the present trend of human thought and start it along new lines completely.

Results? Why, man, I have gotten lots of results! If I find 10,000 ways something won't work, I haven't failed. I am not discouraged, because every wrong attempt discarded is just one more step forward....

Key #3: Exploration

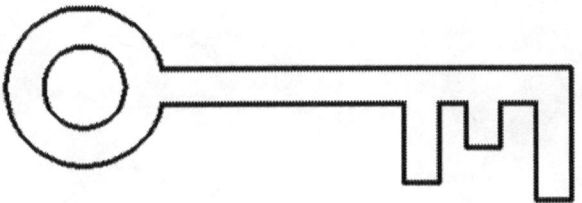

10

Persistence

o o o o o o o o o o o o o o o o o o o o o o o o o o o o o o o o o o o o

Technology is dominated by two types of people: those who understand what they do not manage, and those who manage what they do not understand.

—Putt's Law

When someone mentions the word "exploration," what is the first thing that comes to mind? Is it the exploration of new worlds by historic explorers like Christopher Columbus or Lewis & Clark? Is it space travel, perhaps with unmanned craft heading to Mars? But what about the Bell telephone in 1876? Or the invention of the telegraph in 1837, which Samuel Morse first conceived around 1832? In more recent days, what about the development of object-oriented programming, design patterns, and the expansion of new programming languages that have helped in the exploration and improvement of the technologies driving the Internet? If it were not for diligent exploration, we would not have many of the things we take for granted today.

Research and Development is a critical stage in all facets of design, and is the most common form of exploration in the field of engineering. It is during this period that we pursue the unknown, and to digress into my geeky Star Trek persona for just a moment: "…boldly go where no one has gone before…" Well, perhaps someone was there once, but we're going back for a closer look.

Prototyping a material structure to perform stress tests, creating sample software code, making paper mock-ups of a new building or house, and blowing things up to see how a safety device may work, are not only cool and fun things to do but are necessary steps in finding the strengths and weaknesses of your ideas. Innovations may be polished by practicality but it's exploration that forms the mold.

If done correctly, exploration can also produce artifacts used throughout other future stages of the product development process. For instance, if you decide to test some material for elasticity under temperature, you can pass on these expected limits, as well as the testing process you used, to the Quality Assurance testing teams. Similarly, if you made some software prototypes, you could pass along interface information to developers and those that would be testing the product later on.

Exploration finds more than just the answers we need for the present. It also helps us with solutions for the future. As such, we shouldn't look upon exploration as a needless process that is wasting time en route to market. Instead, we need to incorporate this as a vital step before going on to creation. Even so, exploration can at times be tedious and frustrating. Sometimes the best innovations have the best-kept secrets, secrets that are not easily revealed to the explorer in search of answers. After all, if it were easy, your idea would already exist, for someone else would have it figured out by now.

Exploration has been a cornerstone of human evolution and is deeply engrained into our basic instincts. Exploration has been a part of humankind since our inception. According to many historic theories[1], humankind slowly migrated to various parts of the world until it was virtually all filled up (or at least all the land had someone claiming it). Early man (and woman) set out to explore those things that would guarantee a better way or life, to search for land to sustain themselves with food, shelter, and clothing. Persistence was intrinsic to the dawning age of humankind, as exploration was often a matter of life and death. Perhaps this is why exploration remains deeply rooted in our basic instincts today.

As time went on, humankind explored how to improve upon fire, tools, protection, and even artistic pleasure. In more recent times, humankind has explored things beyond our world, truly going where no one (at least from this planet) has gone before. Humankind has progressed through continual space exploration, as with Russia's first artificial satellite Sputnik in 1957, the Explorer III in 1958 (which discovered Earth's radiation belt), and much later the Space Shuttle, Hubble Telescope, and Mars Pathfinder. We continually explore. It's in our char-

---

1.   i.e. the "Out of Africa", and "Multi-Region" theories

acter to do so. As Frank Borman, an astronaut aboard the Gemini and Apollo missions, once said:

Exploration is really the essence of the human spirit.

How deeply engrained that "spirit" of exploration is within each of us differs greatly from person to person. The level of tolerance, patience, and bravery one can muster up will set that person's course for his or her explorative journey. As with inspirational appetites, the desire to explore and to keep going forward, no matter what obstacles and failures you may encounter, will depend upon your level of exploratory tolerance.

The four keys presented in this book, as mentioned many times, truly intertwine in a symbiotic relationship. If once on the path of exploration we lose our inspiration, then our exploratory tolerance will drop, and we might end up throwing in the proverbial towel. If you should ever find that your inspirational and explorative spirits are becoming weak from failure, and your tolerance is running short, I'd like you to remember something that was said by one of the greatest boxers of all time, the "Manassa Mauler," Jack Dempsey:

A champion is someone who gets up, even when he can't.

That powerful quote has stuck with me for years, and has lifted me out of some deep ruts. No matter how many times I could not find the answers, and no matter how many times the paths of my explorations lead me to brick walls, I remembered that quote, rolled up my sleeves, and went right back at it again.

One story of persistence stands out for me, and that's the story of the light bulb. In the period from 1878 to 1880, Thomas Edison and his associates worked on at least 3,000 different theories to develop an efficient incandescent lamp. Three thousand! Imagine the exploration that they endured. Think back over all the things you have done and been involved in during the past two years. Now imagine that same span of time focused on a single innovation that constantly failed at every turn, 3,000 times. It's a fair amount of time, and takes a lot of persistence.

Edison's innovative lamp would ultimately consist of a filament housed in a glass vacuum bulb, much as we know it today. Although Edison was not the first person in history to attempt an electric type of lamp, he is the man responsible for the first long-lasting bulb using a carbon filament, based on the original patent of Henry Woordward and Mathew Evans in 1875. Arc lamps had also been around for some time, with the most famous arc lamp invention accredited to Humphry Davy in 1809. However, the arc lamp and other designs before Edison's filament bulb were not always practical or long-lasting. This gave Edison inspiration for an innovative answer that would strike him during the exploration.

Edison had his own glass-blowing shed where fragile glass bulbs were carefully crafted for his experiments. Edison was trying to come up with a high-resistance system that would require far less electrical power than was used for earlier arc lamps that were not very practical. This idea of using less power could eventually mean small electric lights suitable for home-use. But finding the answers was not easy.

By January 1879, at his laboratory in Menlo Park, New Jersey, Edison had built his first high-resistance, incandescent electric light. It worked by passing electricity through a thin platinum filament in a glass vacuum bulb, which delayed the filament from melting. Still, the lamp only burned for a few short hours. In order to improve the bulb, Edison needed all the persistence he had learned years before in his basement laboratory. He tested thousands and thousands of other materials to use for the filament. He never gave up. He persisted in this exploration for this new, practical invention.

One day, Edison was sitting in his laboratory absent-mindedly rolling a piece of compressed carbon between his thumb and forefinger. An idea struck, and he went forward. He started testing the carbonized filaments of every plant imaginable, including baywood, boxwood, hickory, cedar, and bamboo. He even contacted biologists, who sent him plant fibers from places in the tropics. Edison acknowledged that the work was tedious and very demanding, especially on his workers helping with the experiments. He always recognized the importance of hard work and determination. "Before I got through," he said, "I tested no fewer than 6,000 vegetable growths, and ransacked the world for the most suitable filament material." His persistence is truly unbelievable.

Edison decided to try a carbonized cotton thread filament. When voltage was applied to the completed bulb, it began to radiate a soft orange glow. Approximately 15 hours later, the filament burned out. Further experimentation produced filaments that could burn longer and longer with each test. By the end of 1880, Edison produced a 16-watt bulb that could last for 1,500 hours, and he began to market his new invention.

Inspiringly referring to his legacy of enduring persistence, Edison was quoted as saying:

> *"The electric light has caused me the greatest amount of study and has required the most elaborate experiments, I was never myself discouraged, or inclined to be hopeless of success."*

Like Edison, you may hit many roadblocks during the R&D process. You're not alone. To quote from a plaque that once hung from the wall of my childhood piano teacher, "A quitter never wins, and a winner never quits."

Keep exploring and remember these three words: persistence, persistence, persistence.

11

Find Your Obstacles Early and Often

Many of life's failures are people who did not realize how close they were to success when they gave up.

—*Thomas A. Edison*

Exploration is not just a matter of finding out what works but is also a process of finding out what *won't* work. Just as important as finding the right ways of doing things is finding the wrong ways. It's quite hard sometimes, being human, to play devil's advocate against ourselves. Nevertheless, this is unavoidable if we are to see our ideas as others will see them. It's better to address all our shortcomings up front than to deal with them later when it may be too late.

As engineers, we strive to make great things. However, at times we tend to make things too great. After all, we're taught to find the best possible answer to a problem. This doesn't mean, however, that the best answer is the most complex one. Instead, the best answer is typically the most pragmatic one. To find the best answer, approach, and design, we need to go in search of obstacles early and often. We need to see the various angles to approach the problem, and really see what will work; not just in theory, but in reality.

Historically, you may never find much information on projects that failed, whether due to poor planning or the lack of constructive exploration. Reason being that history records successful innovations, and tends to forget about ones that didn't make the grade. If a design never came to fruition, then it falls beyond the event horizon of history, and like a black hole in space, is sucked into infinite doom, never to be heard from again.

If you are to have a successful design, and not have your innovative ideas sucked into a void of space and forever lost, you need to approach your exploration knowing there are obstacles out there, and you need to find them. Furthermore, this is not just a one-step process. You need to continually be on the lookout for other obstacles as well. It's best to know your enemy well before going into battle. It's even more crucial to know your enemy well while in battle.

Norman Vincent Peale, author of the famous book "The Power of Positive Thinking" once said:

> *"Stand up to your obstacles and do something about them. You will find that they haven't half the strength you think they have."*

We can't be afraid of the obstacles we encounter. Instead, we need to encourage them, and get them all on the table as soon as possible.

Later on in this book, in Chapter 17, "Post Mortem," we'll be taking a look at an exercise to manage problems encountered after (and during) the creation of a product. Before getting to that stage, however, we need an exercise to manage our obstacles and issues now, during the key of Exploration.

In figure 1, I've illustrated an example spreadsheet similar to the one we'll use later on in the post mortem. Using this tool, you can identify problem areas as you encounter them, get a visual representation of all these obstacles, and manage them in a prioritized manner.

Issue	Contingencies	Priority	Resolutions
Nylon material strength too weak	Will consider kevlar	6	
Compile problem in Unix due to gcc compiler	Possibly use a different compiler for STL issues	4	2/6: Found one compiler from Kai that might work.
100 ohm resistors overheat	Could possibly use more cooling. Can look into better resistors	7	
Steel structure too heavy for enclosure	Look into aluminum or light metals	3	3/1: Found vendor with aluminum extrusions that might help
Under estimated schedule for driver interface	Lengthen schedule, contact stakeholders	0	3/4: Issue resolved. Stakeholders ok'd schedule change.

Figure 1 Issue Management Tool

To use the Issue Management spreadsheet tool effectively, you need to have this setup from the get go. Before you delve deep into your exploration, be prepared with this tool at the ready.

This is a simple tool containing four columns. The first column describes each issue and obstacle. The second column lists the contingencies; the possibilities you could pursue for this obstacle. The third column is the priority of this issue, and the last is a progress log, labeled "Resolutions."

Using this tool is fairly straightforward. As you encounter an obstacle or other issue that comes up in your exploration, list it in the first column. Then give some thought to the various possibilities for resolving this problem and list them in column 2 (Contingencies). There could be more than one contingency to list, so feel free to include as many as you can possibly think of.

Next, rank the priority of the issue or obstacle in column 3. The priority is on a scale of 1 to 10: 1 being a minor issue, and 10 being very important. If you resolve the issue, set the priority to zero.

As you work on the problem, keep track of this in the last column, Resolutions. This column will be a log of your progress. As you add to this column, feel free to adjust the priority column. If the problem gets worse, raise the priority. If the problem is looking better, lower the priority. Once you solve the problem, set the priority to zero.

Lastly, as you work with this tool, sort the columns based on the priority. It's vital that you direct immediate attention to the big issues first, and save your effort for the less significant ones later. It's tempting to solve the low priority issues first since these tend to be the easiest problems to work out. When I think of this, I'm reminded of an old biblical quote, "You strain out the gnats and swallow the camel." This statement was a metaphor referring to how one would strain wine through muslin gauze, ensuring that he didn't accidentally swallow any bugs that may have landed in his drink. The metaphor was to show how people were missing the bigger issues by concentrating too heavily on the smaller ones. The same goes for issue management; don't just strain the little issues and let the bigger ones pass you by. Tackle the big dogs first, as their resolutions may have a greater impact, and could influence the smaller issues as well.

Therefore, to avoid swallowing camels, I like to strain all my issues by sorting them in priority, highest priority first. I like to use Microsoft Excel®, which makes the sorting very straightforward, automated, and intuitive. Sort so that the highest priority items are first. Then, at regular intervals, perhaps once per week, set some time aside to go through the items. You may also find it useful to set deadlines, perhaps in another column, to set goals for overcoming the issues.

Don't hesitate to gather these issues, and review them constantly. It may help to post them in a conspicuous place, perhaps in your office, workshop, or wherever you will usually be when working on your innovation.

Henry Ford is quoted as saying,

> *"Obstacles are those frightful things you see when you take your eyes off your goal."*

Don't take your eyes off your issues. Use an issue management spreadsheet so you can list your obstacles, and manage a plan of attack to deal with them.

In addition to the issue management tool, I use another valuable exercise to deal with obstacles early in the game. I've used this exercise during exploration and during creation as well. I like to call this exercise "The Rotten Egg Hunt." It's a team effort wherein someone will win a prize for finding the most rotten eggs (software bugs, flaws, obstacles, or whatever term you want to use). I found that a gift certificate to Amazon.com works fairly well as the reward. However, you can use whatever you think will help to motivate your team. If the team consists of only you, well, that's a little tougher, but feel free to reward yourself anyway.

The significance of this exercise, no matter if the team is small or large, is to go in exploration of nothing but problems. Take a week and do nothing but the Rotten Egg Hunt. Go in search of as many bugs as you can find. And while doing this, if you're working with a team, foster a competitive theme to the event. If you're involved with a large development or R&D group, perhaps you could choose up sides from which the winning team gets the reward.

The small price paid for the reward to the one finding the most obstacles will be well worth the investment. When you go in search of nothing but obstacles, you tend to focus on anything that can go bad. This gives insight not only to problem areas, but keeps your mind in tune for possible roadblocks later on.

A side effect from finding your obstacles early and often is a continued inspirational spirit, in yourself and in your team (if you have one). By doing a faultfinding exercise early, you're less likely to be hit by curveballs later. You will find that your exploration will progress much more rapidly by avoiding speed bumps later in the game, which could cause severe setbacks. Inspiration breeds success. As you tackle the obstacles early, you will see more and more successes during the key of exploration. This in turn fosters an inspirational environment for continued innovation, and zeal in exploration.

You can also look at early faultfinding another way. Would you rather have someone point out your faults, errors, bugs, etc., or would you rather correct these flaws before anyone has a chance to engage in criticism? Remember that first impressions are lasting impressions. If you bring a product to market, or a design for review, and others poke holes in what you present, not only will *you* lose credibility, so will your idea and your future ideas. If this were to happen, you will not only face an obstacle. You might face the end of the road for what could have been, and what probably was, a very good innovative idea. Hitting the end of the road on a perfectly good idea can really put the kibosh on your inspirational drive, hindering future innovations.

We learn from our mistakes, and learning is a primary purpose of exploration. If we are going into the unknown, to explore the possibilities, then we cannot expect to have all the answers before diving in. Oftentimes, our answers will find us, and they may not be pretty. Nevertheless, we must face our obstacles early and often. Time is paramount in learning from our mistakes, so we need to find these obstacles early. Problems and issues are not just revealed at the initial onslaught of exploration, emphasizing the need to find these issues often. With that said, I will close this chapter with one last quote:

> *There are some things which cannot be learned quickly, and time, which is all we have, must be paid heavily for their acquiring. They are the very simplest things and because it takes a man's life to know them the little new that each man gets from life is very costly and the only heritage he has to leave.—Ernest Hemingway*

12

Innovations Gone Sour

The whole problem with the world is that fools and fanatics are always so certain of themselves, but wiser people so full of doubts.

—Bertrand Russell

Now I am sure that all your ideas are great. And I'm not about to take the time to criticize any innovative thought. Nevertheless, I would like to revisit a few lessons in history, best taught by some patents that, even though they are registered, and had a lot of innovative thought, never really were quite able to pull it off as successful products. These examples help to reiterate many of the topics covered so far, including input from others, being aware of the requirements, knowing the market, and getting a full understanding of all the issues early and often.

As mentioned in the previous chapter, when we talked about early faultfinding, history does not easily reveal projects and ideas that failed. History tends to dwell on successful projects that came to fruition. We hear of success stories like Edison and the light bulb, or Bell and the telephone. These are great success stories of ideas that took the right path. Nevertheless, rarely does engineering or science history tell us of paths to avoid.

I've selected three US Patents to illustrate some valuable lessons we can learn during the key of Exploration. The intention of these lessons is not to mock the patents. Instead, I want to present these to help us steer clear of the path less known, the antithesis of success, the dark passage we wish to avoid: failure. Bear in mind that being US patents, the ideas I'm about to show you were taken through inspiration, innovation, and exploration (or so we would hope). However, these products are not popular today, which begs the question, "What went wrong?"

First, I would like to describe each of these patents, showing you some drawings, abstracts, and descriptions from the patent documents[1]. Then I'd like to discuss some lessons learned from them. I also would like to reiterate that I am not here to criticize these patents, the inventors, or their ideas. In my humble opinion, all ideas all good ideas, but some need more polishing than others. In the case of the three patents illustrated in this chapter, I believe they could use a lot more polishing. Let's take a closer look at them to see what I mean.

The first patent I'd like to discuss is US Patent #6,351,867, the "Body Squeegee."

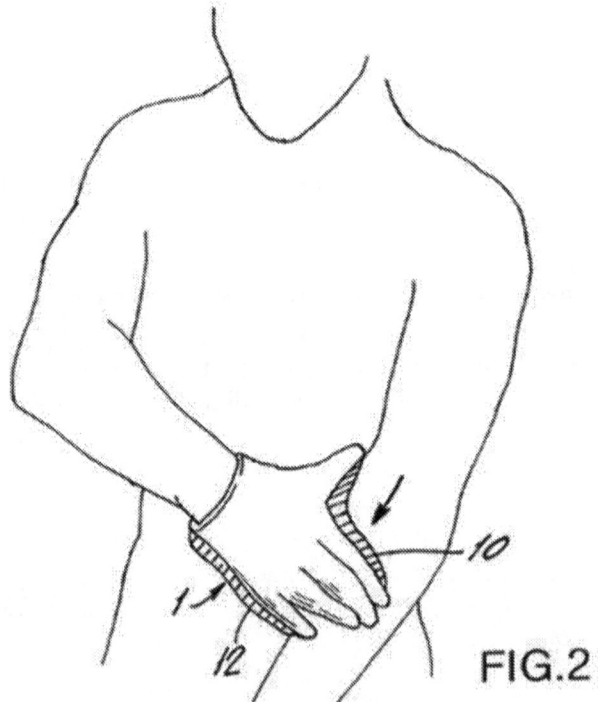

Figure 2 The Body Squeegee

---

1.   All patent information, including images, abstracts, and descriptions are obtained from the United States Patent and Trademark Office which can be found on the Internet at http://www.uspto.gov

The abstract for this ingenious little product goes as follows:

> "A hand wearable body squeegee comprising a glove portion, a concave squeegee band, and a linear squeegee band. The concave squeegee band is formed to contour along the surface area between the forefinger sleeve and the thumb sleeve of the glove portion. The linear squeegee band is formed along the surface area along the pinky sleeve and body towards the opening of the glove portion. The glove portion is constructed from water absorbent material and the squeegee bands are constructed from rubber. An additional feature is that the sleeves each contain a hole for drainage of water. The hole may be located in the area of the sleeve that covers the nail portion of a human hand, or at the end of the sleeve opposite the connection to the body."

That's a lot of words to describe a glove you use to scrape water off yourself when you get out of the tub (or shower, pool, etc.). I guess towels just weren't good enough. This was an innovative idea, but perhaps not very practical. Hats off to the inventor, and don't give up inventing. I'm going to talk more about this patent in a moment, but first, let's take a look at the other two examples.

Next on our list of innovations that missed the mark is US Patent #5356330: Apparatus for simulating a "high five":

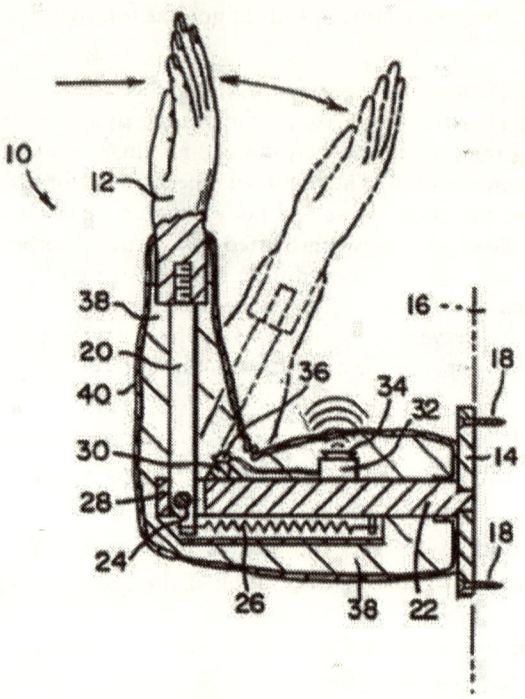

Figure 3 High Five

The abstract on this little goodie points out this thing being an artificial arm that can slap your hand, simulating a "high five."

> "An apparatus for simulating a 'high-five' including a lower arm portion having a simulated hand removably attached thereto, an upper arm portion, an elbow joint for pivotally securing the lower arm portion to the upper arm portion, and a spring biasing element for biasing the upper and lower arm portions towards a predetermined alignment."

The inventor of the high five machine did have some motive for his design. In his patent, he talks about the inspiration behind this innovative idea:

> "During a televised sporting event, a "high five" is commonly shared between fans to express the joy and excitement of a touchdown, home run, game-winning basket, birdie or other positive occurrence. Unfortunately, as known in the art, a 'high five' requires the mutual hand slapping of two participants, wherein a first participant slaps an upraised hand against the elevated hand of

a second participant. As such, a solitary fan is unable to perform a 'high five' to express excitement during a televised sporting event."

More on this in a moment, but first we have one more example to cover.

Although there are dozens of more unique patents from which we can analyze innovations gone awry, let's take a look at one more. This is my favorite, US Patent # 3,241,562, "The Automatic Haircutting Machine":

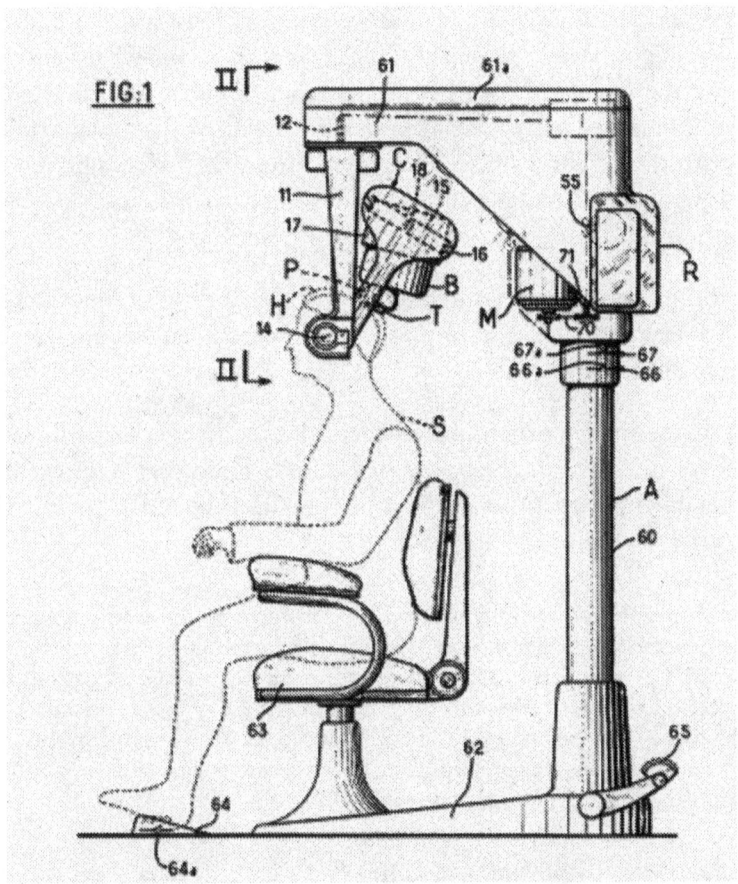

Figure 4 The Automatic Haircut Machine

Although I don't have the flowing locks of my youth, this idea still scares me a bit. Perhaps knowing that there is less distance these days between my scalp and this contraption frightens me even more. Basically, this device would allow you to select the type of haircut you want, at which point you just sit down and it takes over from there. This patent was filed in 1966, and since I haven't seen one of these machines in any of the haircutting establishments that I've patronized over the years, I doubt this idea really took off the ground.

As I mentioned earlier, although I poke fun at these patents, my intent is not malicious. Although I do admit, when looking them over, I smile, squint my eyes a bit, and cock my head slightly to the right in wonder. Nevertheless, each inventor was very serious about what he or she was doing. I just want to point out that I think they might have approached their ideas with a somewhat better methodology, using the four keys described in this book, especially during the exploration stage. These designers were inspired, so much so that they actually went through with the patent process. They were innovative, no doubt about it. But what went wrong? Let's take a closer look at these patents, and see what lessons we can learn from them.

The first thing that pops into my mind when I see all three of these patents is the "next bench" syndrome. I'm sure that these inventors had encouragement from somewhere. But from whom?

Did they invite not only the input of their peers, but of others who could play Devil's Advocate? Take the first patent, the Body Squeegee, as an example. The justification for this invention is described in the Patent document, where the inventor writes:

> "Most people use at least one and typically two or more towels for drying themselves after a daily shower, and it isn't uncommon for athletes, outdoorsmen and physical labors to take two or three showers a day. This results in the collection and laundering of a large quantity of bulky towels on a daily basis, which is a laborious task for today's busy people. The present invention provides a convenient way of drying one's body without the use of towels, thereby decreasing laundering time and making life easier."

I don't know who this inventor interviewed to get this data. I typically use one towel a week, not three a day. Call me disgusting, but it's true. I'd bet a squeegee that most people would find towels more than satisfactory for their after-shower drying needs. And I'd bet two squeegees that this innovation lacked outside input.

Similarly, the next thing that strikes my curiosity about these patents is the market knowledge. What kind of requirement gathering phase lead them to believe there was a market for these ideas? By getting input from not only their peers, and not only engineers, but the actual people and audience this was intended for, they might have seen the market need, or lack thereof.

Let's take a look at the second invention, the "High Five" machine. The inventor mentions that this device is intended for home-use by those watching sporting events. Most likely being a sports fan himself, perhaps the inventor thought he could use one of these. And I'll give the inventor the benefit of the doubt: he may have discussed this idea with some of his friends as well. But what about the larger target market? If this contraption were shown to me, I guarantee my response would not be positive, and I think many others would feel the same. When it comes to supporting my home team, I would rather not rely on camaraderie from an inanimate device.

Sometimes your circle of friends can be very forgiving towards your thoughts and ideas. Good friends will support you, and give you encouragement. However, friends typically don't play Devil's Advocate. If you're shooting for an innovation that goes beyond such a limited demographic as your immediate circle of friends and peers, I suggest you reach out and explore this amongst a larger audience. We'll touch on this point again in more detail in Chapter 17, "Capturing the Audience."

Each of these contraptions was inspired by some kind of thought that lead to an innovative idea for the product. However, how often did the inventors go in search of obstacles? This brings us to the last invention: the haircutting machine. The drawings and 30 pages of patent documents for this machine reflect a lot of work on the part of the inventor. However, by the looks of things, I doubt there were many (if any) prototypes made to prove you could safely cut hair automatically, especially without getting nicked, your hair pulled, or even losing an ear.

We can only speculate as to where these inventions really took a wrong turn, or why they didn't become successful. It's not my place to judge the ideas of others. But I do feel it is important that we study history from all angles; not just the successes, but the ideas that didn't quite make it. I also feel that you should be critical of your own ideas, and continually explore all avenues of your innovation.

Exploration is not just playing with some cool gizmos in your workshop. Exploration is not just downloading the latest software on the market to see what it can do. Exploration is the hunt for the unknown, to seek out what you don't know, and pursue all possibilities.

Get input from your peers. Gather information from your target audience. Look for obstacles early, and often. Always listen. If you are constantly hitting brick walls in your exploration, sit back and evaluate the possibility that you might be in a stalemate. If so, perhaps it's not wise with this particular innovation to pursue the next level, creation (at least not yet). Instead, take a few steps back and cover the possibilities. Get more innovative input, strengthen your inspiration with the advice from others, realize your requirements, and explore some more.

13

To the Moon

Don't tell me that man doesn't belong out there. Man belongs wher-
ever he wants to go—and he'll do plenty well when he gets there.

—*Wernher von Braun, in 'Time' magazine 1958*

Space exploration has always amazed me. I recall a time back in 1993, while working as a mechanical engineer for a company contracted to do work at Edwards Air Force Base in Southern California, that I had the rare opportunity to watch the Space Shuttle land. For me, it was a once in a lifetime opportunity, mostly from just being in the right place at the right time. Though most civilians like me have to remain in an observation deck hundreds of yards away to watch such events, the guards granted me permission to walk onto the tarmac with the other Edwards' employees to watch the shuttle touch down.

As dual sonic booms thundered overhead, I watched in awe. A speck of pepper from the blue sky grew larger as it approached. As the shuttle drew near and landed in front of us on that runway, I'll never forget how fast my heart was pounding. That such a marvel of space exploration was landing right in front of me, though moments before it was outside the realm of planet Earth, astounded me and filled me with fascination.

When I think of exploration, I'm reminded of the accomplishments and bravery of our country's space program. Exploration in the space program is not just a matter of going beyond our planet to discover new frontiers. It's also the ground crews and the engineers who, through painstaking efforts, innovate and explore how to perfect the crafts that launch into space.

Out of all the great stories regarding the exploration of our space program, for me one story stands out above all the rest. It's a story of inspiration, innovation, and exploration; a story that to this day brings me encouragement. It's the story of a man, Homer Hickam, who against all odds pursued his boyhood dreams. It's from Homer's story that we can gain insight into the essence of Exploration, and see what it sometimes takes to persist through this key with its progenitors Inspiration and Innovation.

The movie "October Sky" tells the story of Homer Hickam's passion for the exploration of space, based on Homer's true-life adventures from his book, "The Rocket Boys." I thought it fitting that before moving on to the last of the four keys, we close out the topic of Exploration with Homer's inspiring story of innovative exploration. And, if you've never seen the movie, I highly recommend it.

As the story goes, Homer was living in a dead end coalmining town, with no apparent future, when even the future of the world was in doubt. The year was 1957, and the communists had launched Sputnik into orbit.

According to the movie, Homer Hickam (played by Jake Gyllenhaal) grew up in Coalwood, West Virginia. The Olga Coal Company owned the mountain, the town, the land, the houses, and the lives of all the people who resided in Coalwood. John Hickam, Homer's dad (played by Chris Cooper), was the superintendent of the coalmine. John *lived* for the Olga Coal Company and showed pride in his elder son, a high school football star, but showed far less interest in Homer, the younger of the two boys.

Inspiration was hard to come by in Coalwood, where a man's life was typically destined for the mines. Nevertheless, that didn't stop Homer from staying inspired to pursue what he yearned for. Without the encouragement of his father, combined with a dead end outlook in a dead end town, Homer astonishingly found his inspiration elsewhere.

When Sputnik flew overhead one night, a tiny bright light in the predawn sky, Homer's world opened wide, as though he had seen his future for the first time. Inspiration began to flow, and Homer, not knowing it at the time, was about to embark on an amazing journey through innovation and exploration.

Making an unlikely acquaintance with a distant, geeky, yet intelligent boy named Quentin (played by Chris Owen), Homer went on a pursuit for knowledge. Homer and Quentin teamed up with some of Homer's closest friends, who also shared the common interest in this new world of space travel and rocket development. Together this team, coined the "rocket boys," ventured on a quest of scientific exploration.

As the story goes, the Rocket Boys created a variety of amateur rockets, inspired by this new era of space exploration. But time and again, their experiments ended in failure. Numerous attempts would fail in on-the-ground explosions, or a misdirected rocket gone out of control. Failure after failure, the boys persisted to get it right.

Homer and the other "rocket boys" did everything they possibly could to finance their rocket building activities, which wasn't easy while living in a poor West Virginia coal-mining town. At every step, Homer and his dad fought about the kid's "foolishness." Nevertheless, the "rocket boys" persevered, and continued to pursue their exploration of rocketry.

It was inspiration that stoked the burning desire of Homer's technical pursuit, despite a rough battle through innovative exploration. Chronicles of Wernher von Braun, one of the world's first and foremost rocket engineers and a leading authority on space travel at the time, inspired Homer. Homer looked upon Wernher von Braun with admiration and respect, and dreamed that one day he too would be involved with rocket science. Additionally, Homer and the rocket boys had the encouragement of their teacher, Freida Riley, who gave them moral support and helped them set goals for their scientific exploits.

As time went by, the "rocket boys" had successes, which further inspired them throughout their lives. A poignant part of the story is when the boys do reach success, and are in control of their designs. Their small rockets were now more predictable, resulting from extensive exploration, and acceptance of other failed attempts.

The inspiration Homer got from those days as one of the Rocket Boys stayed with him throughout this life, and led to greater accomplishments. Homer graduated from Big Creek High School in 1960 and from the Virginia Polytechnic Institute in 1964 with a BS degree in Industrial Engineering. A U.S. Army veteran, Homer served as a First Lieutenant in the Fourth Infantry Division in Vietnam in 1967–1968, where he won the Army Commendation and Bronze Star medals. He served six years on active duty, leaving the service with the rank of Captain.

Homer then went to work as an engineer for the U.S. Army Missile Command from 1971 to 1981, assigned to Huntsville, Alabama, and Germany. In 1981, he went to work as an aerospace engineer for NASA at Marshall Space Flight Center.

Some of Homer's accomplishments since those early "rocket boy" days include awards such as the Astronaut Office's coveted Silver Snoopy award for his outstanding support of the astronaut corps, and a special commendation for overall excellence from the Director of the Marshall Space Flight Center. His specialties at NASA included training astronauts on science payloads, and extravehicular activities (EVA). He also trained astronaut crews for many Spacelab and Space Shuttle missions, including the Hubble Space Telescope deployment mission, the first two Hubble repair missions, Spacelab-J (the first Japanese astronauts), and the Solar Max repair mission.

Homer is a textbook example of someone who had an innate, natural sense of the four keys. Great scientists and teachers and a hunger for the exploration of space inspired him. He was vigilant in his innovations with rudimentary rocket building, and his limited resources from which to work. Exploration was a never-ending process, from which many great things came as a result.

In turn, Homer serves as an inspiration to others, especially on that sometimes-difficult road of exploration. To this day, I get inspired by Homer and his persistence in exploration.

Homer's story shows us how the key of Exploration is highly dependent on its lower predecessors, Inspiration and Innovation. During what can be a trying time throughout the key of Exploration, we need persistence to stay inspired and continue to innovate. Homer had an abundance of inspiration. From an early age, he was able to find his path through introspection, looking inside himself and pursuing his dreams. Staying on this course—to do what he yearned for in life—nothing could stand in his way. Nothing could deplete his inspirational appetite. This strong, persistent inspiration is what led Homer through innovation, exploration, and to great things throughout his life.

Don't let obstacles stand in your way. If, during the key of Exploration, you find that your innovations don't blast off, stick with it and keep trying. Remember that some of your innovations may temporarily fizzle before leaving the launch pad and finally heading for the stars.

Be persistent, and don't give up easily. If you stay inspired and continue to innovate throughout the key of Exploration, you will find a clear-cut path to the next of the four keys: Creation.

Key #4: Creation

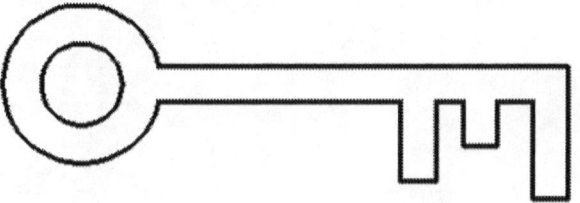

14

Waterfalls and Spirals

o o o o o o o o o o o o o o o o o o o o o o o o o o o o o o o o o o o

Good judgment comes from experience, and experience comes from bad judgment.

—Barry LePatner

Creation, the last of the four keys, is not a one-step formula. Inspired innovative thought has led us through so much exploration that it seems like we could just take one last step, and create that which we originally envisioned. However, that dive-in-head-first approach would likely lead us to failure. We've come so far with our exploration and peer reviews, and have possibly had to go back to the proverbial drawing board more than once. At this point, it's more important than ever that we tackle creation carefully, cautiously, and—as I'll explain in this chapter—iteratively.

In this chapter, we'll digress momentarily back to "Process 101," or some similar class you may have covered in school. If you never had the opportunity to take such a course, not to worry: we'll cover it in brief so you can see the benefit of the waterfalls and spirals presented here. If you thought this chapter was about amusement park rides, I hate to disappoint. But not to worry, and don't let the term "process" scare you either. The fourth key, Creation, like the other three keys, is a methodology and approach. Creation is a way of getting your inspired innovation from exploration to fruition.

Creation is not just a process. Creation is a key that feeds off its lower predecessors: Inspiration, Innovation, and Exploration. As with the other three keys, Creation relies heavily on the symbiotic relationship between all four keys. Creation can be a difficult step to tackle, requiring our Inspiration more than ever. Things will not always be straightforward as we pursue this last of the four keys, making the key of Innovation vital to its success. And as we'll see throughout this chapter, and those that follow, Exploration will be ongoing.

During the key of Creation, we'll take a look at three standard engineering processes and see how the keys of Inspiration, Innovation, and Exploration play a vital role. I know that term "process" makes it sound like we'll be entering into the humdrum of procedural development or project management. But we won't. I don't want to re-teach what other books and college courses already have. Our goal here is to see how the key of Creation, and its interaction with the other three keys, works into some standard engineering practices, which you may have already learned. If you've never heard of these processes, we'll describe them briefly here to understand their basics. The approach in this chapter takes us beyond just knowing the steps. We'll see how the interwoven relationship of the four keys will help us to finally bring our ideas into reality.

As you may recall from the topics in the "Exploration" section of this book, we went through somewhat of an iterative process. One of the principle fundamentals of exploration is to keep getting feedback, and try things as we move forward, thus allowing ourselves to experiment with baby steps before deciding to run the gauntlet. During the key of Exploration, I emphasized the importance of finding your obstacles early and often. And when discussing the key of Innovation, I pointed out the importance of getting input from peers and others on a continual basis. During Creation, we also need to take many smaller steps and "iterate" through this key.

Now that we are ready to create something (or revise an existing product), it's advisable to maintain the same incremental philosophy previously discussed. Covering this methodology, we'll see how three well-known industry standard processes relate to successful and motivational creation, and how to best use these processes to avoid having your tower of developmental success tumble and fall.

Each of the three processes discussed in this chapter deal with some standard steps to make a product[1]. Some of the more noticeable steps in each of the three processes are:

---

1.   In this context, I refer to the term "product" as the end result of your idea. This could be a manufactured product, house, clothing, etc., whatever your initial idea was.

- Requirements gathering
- Analysis and design
- Implementation
- Test
- Deployment

For clarity, in the context of this chapter, I'll be using only these five process steps. In many cases, however, other steps need consideration, such as Business Modeling and Maintenance. Nevertheless, to keep it simple, we'll just concentrate primarily on these five.

All of these steps during the creation phase can be either addressed one at a time, gone through little by little, or we could attack each one at various times with varying weight. The three processes in this chapter describe just that.

The first of these processes is known as the Waterfall Process, shown in figure 5 below.

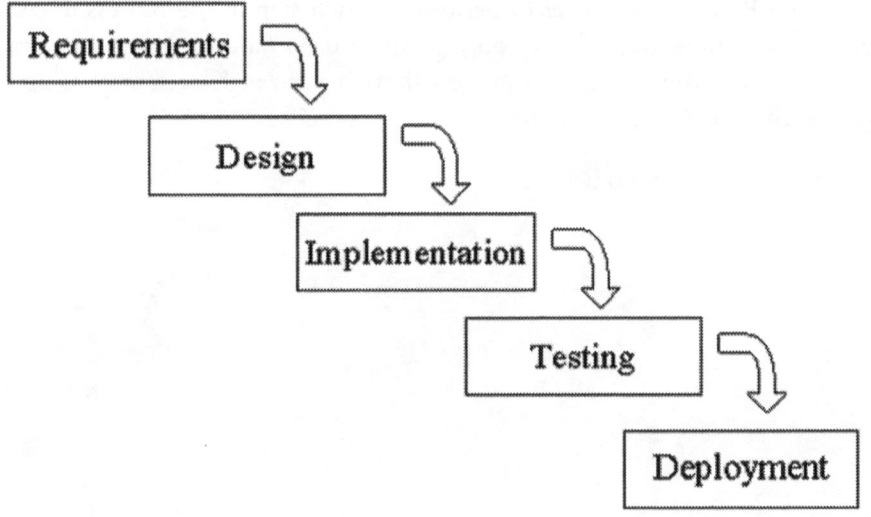

Figure 5 Waterfall Process

This is probably the most familiar process among engineers, as it's one of the oldest. In the Waterfall Process, we take a step, finish it completely, and then move on to the next step. However, as pointed out in the "Exploration" key, this forward-only approach tends to lean towards failure. By doing all the design and then all the implementation before testing a product, we may have left out valuable input from others (including those testing it) and we might not have performed enough exploration (perhaps prototyping) before committing to an implementation phase.

Even though the waterfall process is well-known, it's rarely used today, especially in software design. However, for scheduling purposes, program management departments tend to use this as a blanket to cover smaller sub processes, which usually contain one of the next two processes I would like to talk about, the Spiral Process and the Unified Process (UP).

Both the Spiral Process and Unified Process are known as iterative processes. During these processes, you will be digging into each step a little at a time, iterating until it's complete. Not only do these processes lend themselves well and efficiently to the key of creation, but as we'll see, they are closely tied to the other three keys. Let's take a closer look.

The Spiral Process is a matter of performing each step in the process once, but not to completion. Instead, you only go in so deep (and in many cases, rather shallow). Then you repeat this process, iterating through each step again and again, each time doing a little more.

This is shown in figure 6 below.

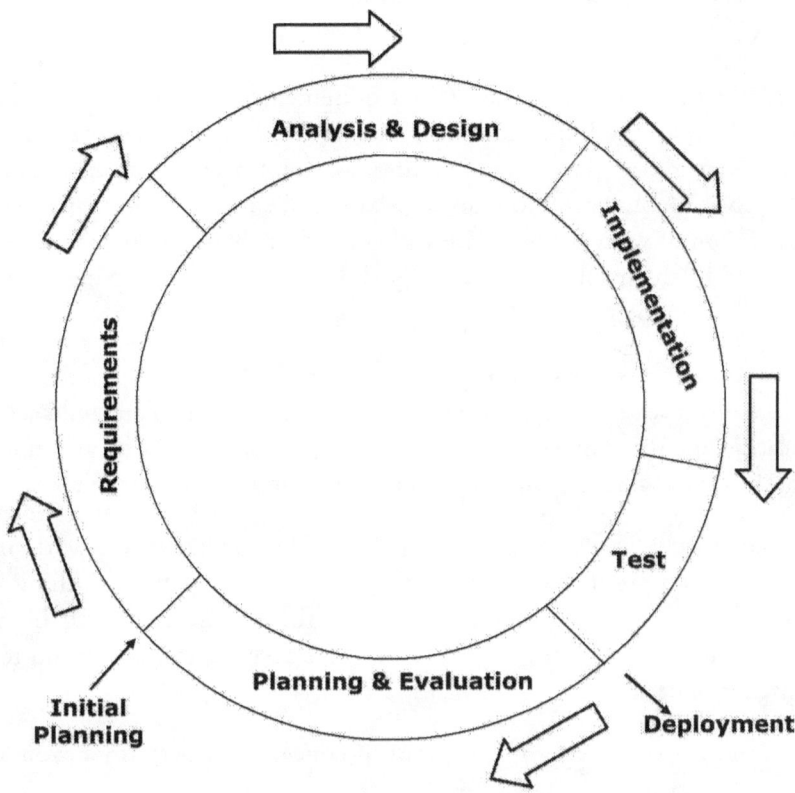

Figure 6 Spiral Process

The idea of the Spiral Process is to take a baby step through each of the process steps continually until you reach a full and final deployment. After each iteration, there could be a single, smaller deployment. If you were creating software, each iteration may only complete 25 percent of its features.

For instance, let's say you are making a web site. You (or someone) had an innovative idea for this site, and created some initial planning based on some of your exploration. You know this site will be used to sell commerce, perhaps shoes. Your exploration, before hitting the key of Creation, shows you will use a database and perhaps ASP as your primary scripting source. You had time to fiddle around with these things during the key of Exploration. Now you are satisfied that they will bring your inspired innovation to reality. It's time to enter the Spiral Process.

The first thing you do is gather as many requirements as possible. This could be a matter of defining the limits and scalability of the web site, and what it should look like. Remember, since we're iterating, we don't need all the answers on the first iteration. We just need to know the basics. We need to make sure we cover the foundation of what this web site will be, and move on. Some of the requirements could be the maximum number of shoe styles, and the maximum concurrent sales that can be processed.

Then you do some analysis and high-level design. Perhaps you see if the database you selected can scale to the requirements you just gathered. You could see if the specification for the database software lists the scalability limits, and how that compares to the maximum number of shoe styles and concurrent sales.

Next, you do some implementation, not all of it, but perhaps some working prototypes to show a proof of concept on how the site will operate. Nothing fancy, but something to prove your ideas based on the requirements and high-level design. For instance, you might have a web page access the database and retrieve some pseudo shoe styles.

Then you test it out, seeing if this proof of concept will perform as you anticipated.

In the last step of this first iteration, you regroup and check your planning to make sure you are still on track. It's time to break out the compass and map, and make sure you're still on course. Will it take longer to get the database online? Do you need a different database altogether? Did you discover that the ASP scripting might take longer? Review your steps in this iteration, and see where you are.

It may be obvious at this point how the key of Exploration played a role in our first iteration, primarily during the Analysis & Design step. What might not be so obvious, however, is that the keys of Inspiration and Innovation play a very important role as well. I hate to use the term "process" during the key of Creation but we need to have structure and steps to follow. Nevertheless, there is also a human element at work here.

As I mentioned earlier in Chapter 1, inspiration is often overlooked when pursuing an innovation and going through the procedures that bring it to fruition. As engineers, designers, and architects, we are taught the fundamental laws of science, and the events required to design a product. But without inspiration, our momentum at any stage of the game, including the key of Creation, could diminish. Furthermore, should we have hit any obstacles along the way during an iteration, we need to be innovative to overcome them. Without our inspiration running at an all-time high, our innovation will not be as effective.

By completing our first iteration, we're now able to see something blossom. Early on, we planted a seed with an inspirationally motivated innovative idea, then watered and nurtured it through the key of Exploration. Bringing this sapling to the key of Creation, we spiral through our first iteration, and now we see something start to bud and blossom. By taking this first small iteration, we begin to see the fruits of our labor. Like a proud father or mother, you can see the first signs of life in the womb. If, on the other hand, we would have taken the waterfall approach, we may not have seen any life signs yet, discouraging our inspirational drive.

Now that the four keys have fueled the momentum in our last key of Creation, it's time to tackle the 2nd iteration. This follows all the same steps as the 1st iteration, but this time it's a little different. This time you have a proof of concept, and a better understanding of the design. Still, you revisit the requirements, revisit and refine the design, then implement the proof of concept a bit further into more of a *real* product. Then of course you test what you have, and revisit your planning and scheduling.

You continue through a number of iterations until you're finished. At one point along the way, you may have an iteration complete enough for a beta product to be field-tested, or perhaps an early release. In either case, using this spiral approach, you can take baby steps, and see your obstacles early and often, continually improving your idea. In the case of an iteration producing a beta or early release product, you might be able to start a revenue flow before the final product has completed all remaining iterations. That would definitely be inspiring, and help fuel the key of Creation even further.

This leads us into the Unified Process (UP) shown in figure 7 below.

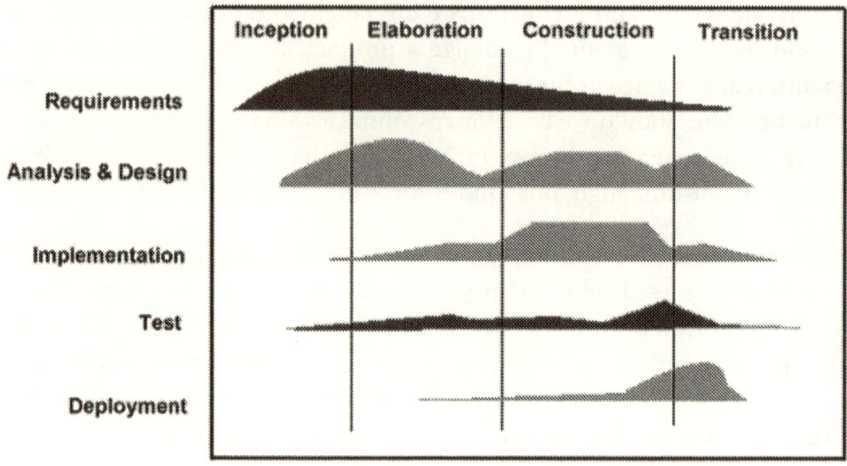

Figure 7 Example of the Unified Process

UP is by far my highest recommended process for the creation phase. This process is indeed a spiral type of process, and has four primary iterations[2]: Inception, Elaboration, Construction, and Transition. Each of these iterations is like a baby step (though slightly bigger, the concept remains the same). As shown in the UP diagram, all the elements that participate in the process (Requirement Gathering, Design, Implementation, Test, and Deployment) participate through each iteration. In other words, we spiral through each of the elements that comprise the entire process.

The unique thing about UP, however, is that it gives weight to the amount of participation that each element should have at a given iteration. For instance, looking at the first iteration (Inception) we can see that there should be more concentration on Requirements than on Test. And note that there is more emphasis on Design than on Implementation during the first and second iterations. As we approach the 3rd iteration, Construction, there is a lot more emphasis placed on Implementation than on Requirements or Design.

---

2.   The UP diagram shown here is a simplified rendition. UP often refers to Inception, Elaboration, Construction, and Transition as phases with each containing many iterations. For more detailed information on UP, please see the Recommended Reading section near the end of this book.

Notice also how UP places Deployment, actually getting something in the hands of your customer. UP places Deployment as early as the 2nd iteration (Elaboration) and heavily in the 3rd iteration. This emphasizes a crucial point of finding your obstacles early—discussed earlier in the Exploration key—and a point discussed in Innovation concerning Next Bench, and getting input from others early on. UP drives home the fact that the customer is the ultimate designer of our product. We need to build what they want or need. As such, we need their input early and often.

It is through this ingenious UP process that we can maintain all four of the keys presented in this book. We're at the stage where we've been inspired, had some innovative thought, and explored the possibilities. Now we need to:

1. Move forward with solidifying our requirements to ensure we've set our stakes in the ground.

2. Perform analysis and design to get this thing on paper and aligned with all aspects of construction and development.

3. Start making what we can to start the realization of the idea.

4. Test this puppy to see if it barks.

5. Deploy it.

As mentioned earlier, the waterfall process takes this approach with steps that flow only one way: down. Just like Niagara Falls, there is no turning back in a waterfall process. There is no such thing as testing the waters before hopping in the barrel and taking a plunge over the edge. I prefer a much stronger safety net than that. Using an iterative process, I get just that: the safety and assurance during the key of creation.

Spiral and UP are very similar. These allow you to test the waters beforehand. This allows you to see any obstacles early on, continue the involvement of all the stakeholders and participants, and still take baby steps so you can easily take one or more steps back if need be, before it's too late.

I'd also like to point out that the Spiral and UP processes are not cast in stone. Each idea you have, and every project you work on, will vary. Similarly, you need to vary the process to fit that particular need. Every project where I've used Spiral/UP has been a custom derivation, but maintains the same basic principles of iterative processing.

The Spiral Process and Unified Process are sciences unto themselves. To explain these processes in detail could fill another book (or two). I've only scratched the surface of this topic and I encourage you to explore these processes further. Please see the "Recommended Reading" section near the end of this book for my recommended references on this subject.

Creation thrives on inspiration. As with the Spiral Processes discussed in this chapter, we need to place iterative weight on the other three keys when tackling Creation. We need to stay inspired, and never lose hope if we hit an obstacle. Innovative thought is crucial, and at an all-time high at this point as well. We need to continue questioning ourselves, and proving that this idea will work. Once again we see how closely related these four keys are: Inspiration, Innovation, Exploration, and Creation. Even during the final phase, we cannot abandon any of the four key principals that brought us to this point.

15

Flying Cows and Monuments

o o o o o o o o o o o o o o o o o o o o o o o o o o o o o o o o o o
If a customer asks for a glass of milk, don't design him a dairy farm with flying cows.

—*Nathan Cool*

I came up with that quote one day while sitting with my brother Dan. We were discussing a recent project I had worked on wherein things went awry. It was supposed to be a simple software project, but the more engineers the team gained, the more it resembled the tower of Babel. I could see that during the key of Creation we were not addressing the customer's requirements, and missing the market. The customer didn't need such an elaborate design. They needed something that worked, and worked well. Metaphorically speaking, the customer was asking for a glass of milk, and he wanted it now. Nevertheless, the project team, guided by majority rule, decided to build something more scalable, more flexible, something bigger and consequently bulkier, a dairy farm with flying cows.

I had seen it before, and I was seeing it again. A design approached with good intentions to make a product better than expected with a goal to satisfy speculated future requirements. All that the customer wanted was a simple glass of milk. But the team set out to make this a *perfect* design. This perfect design would satisfy even the smallest of requirements, some of which the customer wasn't even aware of yet.

After all, how could the customer think he only needs a glass of milk? As engineers, the team felt they should look out for the customer's hitherto unseen requirements. To do so, they would build something more scalable than a glass of milk: an entire dairy farm. Better yet, just to be on the safe side, and to make this better than the competition, put some wings on the cows just in case.

We were making software that had nothing to do with agriculture. However, the analogy just seemed to fit. I did not feel comfortable with the project and I eventually left before things got worse. The reason I left is the subject of this chapter. I've learned in the past, the hard way, a lesson that many engineers have also faced. This lesson is simple.

A perfect design is an enemy of a good design.

Oftentimes, engineers striving for a perfect design may end up with no design at all. If it takes too long to design, and costs too much to make and maintain, then you could completely miss your market. A simple design may not provide the best and ultimate solution to a given problem, but it might have the best chance of meeting the schedule and cost constraints with acceptable quality. Additionally, a simple design is easier to implement, maintain, and enhance.

If you'll recall from Chapter 7 when we discussed cognitive distortion, the world is not made of perfect things. Instead, we live in a world of *nearly* perfect things. Although we should strive for perfection, we need to be pragmatic at the same time.

Well, the other engineers on that winged bovine project learned their lesson the hard way after the company doled out millions of dollars to find out they were on the wrong track. The project encountered inspiration's worst nemesis: failure. Like a disease, this spread from that small engineering group to other departments. This had a domino affect on the other keys, such as Inspiration and Innovation, that were essential to the recovery of this project and future projects as well.

So how do you avoid this pitfall? First, understand that there is a difference between "flying cows" and (what I call) monuments. There are times when a design requires the highest of sophistication. Take the Boeing 747s or the Space Shuttle as examples. Compromise design on these projects and you jeopardize people's lives, *and* risk millions or billions of dollars. These are what I like to refer to as monuments. Designs like the large aircraft we fly in today, bridges, skyscrapers, and spacecraft are *monumental* designs. They pay tribute to the innovation of humankind and mark history with exceptional achievement. When it comes to monuments, it's more difficult to fall victim to the flying cow syndrome, but it is still possible.

Not all designs can be simple. However, all designs must be effective. Furthermore, many designs can be simple. There are times that you need monuments. However, make sure that is what your market is asking for. Do not turn monuments or other requirements into flying cows.

If you are designing something that is going into space, or is life critical, then you need to be careful on simplicity. Nevertheless, there is a practical approach to avoid the dairy farm syndrome. During your requirement-gathering phase, before entering the key of Creation, as well as during each of its iterations, make sure you get three crucial items specified:

1. The Product Life Cycle.

2. Budget.

3. Time To Market.

All three of these elements are closely related and intertwined. If you vary one of these elements, the other two could reciprocate or proportionally distort. It's definitely a balancing act to keep all three in check. Let's take a closer look.

The Product Life Cycle defines how long your product will theoretically exist. If you are making a house, then this cycle will be quite lengthy, perhaps 20, 30, or 50 years, or more. However, if you are making a toaster, then the expected life may only be two or three years (or that is all you want to shoot for). If you built a toaster designed to last more than 50 years, it would most likely turn out to be far too expensive and un-sellable to your market.

Additionally, if you are making a web site, the Product Life Cycle may be less than 1 or 2 years. After all, things can change rather quickly in the world of the web. In either case, you need to know the length of time that your design will have to sustain. This will help you to decide how much infrastructure needs to be in place for maintenance, and structural strength for longevity.

Budget is critical, as this will help you determine not only the materials or components you need to design, but also to what quality, and time invested. Never forget that time is money, meaning that time is a crucial part of the budget to analyze. Every innovation has an expected return of investment. You need to ensure that your budget is strictly maintained to avoid overrun costs, which decrease the financial return.

Many companies rank the importance of time-to-market very high. Without a product in the hands of the customer, there is no cash flow, which means there is no company. Being the first on the market with an innovation can also get you the most attention from prospective customers. This tends to give you an edge over the competition, not just in the short term, but for long-term customer loyalty as well.

As mentioned, all three of these things, Product Life Cycle, Budget, and Time to Market, are closely related. If you have a low budget, then your product life cycle may have to sacrifice, and be short. However, this could allow a faster time to market.

If the budget is high, then you might be able to lengthen the product life cycle by building a better, more stable product; a product with higher quality and standards. However, you might hinder time to market since it could take longer for the design enhancements.

In any case, knowing the three elements of Product Life Cycle, Budget, and Time To Market, you can analyze your requirements to avoid making flying cows, while at the same time making monuments where need be.

Remember that there is no such thing as a "perfect" design. Strive for perfection, but while you do, maintain practicality and balance the three elements that avoid the flying cow syndrome during the key of Creation.

16

Capturing the Audience

o o o o o o o o o o o o o o o o o o o o o o o o o o o o o o o o o o o o

"It would be possible to describe everything scientifically, but it would make no sense; it would be without meaning, as if you described a Beethoven symphony as a variation of wave pressure."

—Albert Einstein

Our innovative ideas will one day, if all goes well, be in the hands of those who will make use of the final product: the customer. If you're designing a house, this customer is the homeowner. If you are developing medicine, it's the patient. If you are designing software, it could be a number of different people, from consumers at home to specialists for whom your program was designed. In any case, we need to keep our innovations customer-driven, using our customers as the ultimate designers.

Remember that if your customers can't use your products because they can't relate to them or figure out how to use them, or find them annoying or far too expensive, then the customers won't buy your products and may speak ill of them as well. During the key of Creation, it's important that we involve our audience. Similar to dealing with obstacles discussed in Chapter 11, we need to involve our audience early and often.

A few years ago, I bought my parents their first computer. My folks were in their early 70s at the time. Living far from our kin, we wanted to keep our parents connected to their kids via e-mail. My folks lived their lives in a different era than their younger kids. To them, the computer was a mysterious machine that looked liked a cross between a typewriter and a television. This new computer was a learning experience for not only my parents. This was also my first true eye-opening realization of the term "metaphor" as it applies to a product.

Metaphors are things conceived to represent something else. Software probably has the most intense issue with metaphors, as it needs to represent something unnatural in a natural way. For instance, how do you represent a file on a computer? Inside the computer, a file is a collection of bytes, typically stored on a disk drive. In the real world, however, a file is contained in a manila envelope, and most often stored in a file cabinet. As a graphical aid, computers tend to use a metaphor of a manila file folder as an icon to represent a file on the computer; thus providing a natural representation of an unnatural thing.

After my parents received their first computer, they were sharp enough to get it running. But try to explain what a "window[1]" is. They kept calling them boxes. They thought the window was the entire screen; after all, windows are made of glass, right? Sometimes they would refer to the screen as the picture tube, applying a television metaphor.

I would explain how they could minimize their "boxes" on the "picture tube," but the icons were foreign to them and made no sense. The mouse was another challenge. The Windows® ME operating system installed on my parents PC was easy enough for them to learn, with a little coaching from my brothers and myself. Microsoft has come a long way in improving the metaphors in their software, and I'm not intending to knock them by any means. Nevertheless, with all due respect to Microsoft, it was a revelation in the world of metaphors, sitting my 70-plus-year-old parents in front of their first computer, even with such an easy-to-use, de facto standard operating system like Microsoft Windows®.

Even the simplest of things for us, like Windows®[2], might be difficult for someone else. During creation, there is a natural tendency to concentrate heavily on the internals of our innovations, and not the actual representation. We need to bear in mind the metaphors we are trying to represent. This holds true not just for software, but for all aspects of engineering as well. There may be no perfect metaphor for what you are making, but try to get as close as possible. The more that people find your product easy to understand, the more they will appreciate it, get use out of it, and come back for more.

---

1.   My parent's computer had Microsoft Windows® ME software installed.
2.   My apologies to MAC and Redhat users. Those are great OS's as well.

Let's say you are designing an automatic window opener for use in the home. The idea behind your design is that an electric motor will slide the window in a room open or shut when the user activates this automated mechanism, perhaps using a switch or other activation device. The user needs to activate this open-close mechanism, and he or she needs to intuitively understand how to access this activation device. So what metaphor would you present to the user? Would it be a drawstring-type cord like the ones on mini blinds? Would it be a wall switch like the ones used for turning on lights? Or perhaps a dial that when turned clockwise would open the window, and when turned counterclockwise would close it.

No matter what your design, you always need to keep the target audience in mind throughout the key of Creation. Their input will not only drive your metaphors, but also shed light on some aspects you may have overlooked.

I've found that focus groups work well as a means of listening to and capturing the feedback from your audience. In a focus group, you bring together some users, perhaps five users at a time, to discuss issues and concerns about the features of your product. You (or better yet someone impartial) work(s) as a moderator, maintaining the group's focus and running the exercise.

Participants in the focus group should feel free and relaxed. These people need to feel as though the exercise is relatively unstructured, but in reality, the moderator needs to follow a specific preplanned script of issues and set goals for the type of information to be gathered.

Using the automatic window opener idea, you could organize a focus group of some homeowners. Make a prototype of the window; perhaps it doesn't even have a motor attached to it yet. Then, install one of the activation ideas, perhaps the dial, on the wall. You might make a script that goes something like this:

1. Before entering the room, explain the window opener concept, but not how to activate it.

2. Bring one (or all) user(s) into the room.

3. Ask user(s) to open the window using the new device, but don't point to the activator.

4. See how many users go to the window in search of the activation device, and see how many look at the wall for a switch or dial.

This tiny script could quickly reveal how many people prefer the drawstring metaphor on the window to the dial (or switch) idea on a wall, or vice versa. If more people walked to the window, then perhaps the better metaphor would be to place some type of drawstring-activated device near the window. On the other hand, if the users were looking at the wall, searching for a switch or dial, then this might be the more appropriate metaphor.

The importance of capturing the audience is to do it early and, like with our previous search for obstacles, do it often. Referring back to the Spiral Process in Chapter 14, we could perhaps organize a focus group during the second iteration, or perhaps right after the first iteration[3]. The sooner we get feedback from our audience, the sooner we can solidify the requirements.

You could even spiral further into this study by having additional focus groups following the first. Going back to our automatic window example, let's say you determine in the first focus group that users went to the wall in search of the activation device. You might have an additional focus group with a new, fresh set of users to see if the switch or the dial would work better. In this example, you determined that users seem to prefer a wall-activated device. The second focus group drilled deeper to see which one worked best (switch or dial).

Bear in mind that reading human actions are not all that is required from focus groups. After your initial tests, ask for input from the users. Let them freely tell you what they liked or didn't like, and get their ideas.

Focus groups are merely one way of getting empirical data concerning your product. Another approach is with beta products. This is a lot easier when it comes to software, but it could apply to almost any product you are designing. For instance, say you were designing a new type of outdoor clothing. The material for this clothing would be waterproof, keep you warm if it is cold, and keep you cool if the weather is hot. The material would also be flame-resistant, comfortable, lightweight, and resistant to insect bites.

For this product, you might think about making some prototypes during the creation phase, and perform tests on the material. But this is only proving the concept in the lab, and you don't have empirical data from the audience. Getting user data, you can truly find out such things as comfort, desire for colors, and if

---

3.   Recall from the UP process discussed in the "Waterfalls and Spirals" chapter that Deployment starts as early as Elaboration, very early on.

the product will really live up to its claims in the real world. An option for tackling these issues is to make a beta product, one that is perhaps hand sewn, and maybe only the shirt or just the pants. Then, by having some users try this out, you may get other feedback such as:

- There are not enough pockets.

- Use zippers instead of buttons.

- The insect repellant in the material works, but after a few hours the smell makes you nauseous.

- The color you designed works well in the California desert, but in Alaska it tends to attract grizzly bears in mating season.

The point is, you really don't know what to expect until you get the product in the hands of the customer. They'll find things you wouldn't expect. I ran into an example of this some time ago when we designed a new chassis to house some communications equipment. We spent a lot of time getting the features into the equipment, and the software to drive it. Our primary focus was on the stuff inside the box, and unfortunately, we paid far less attention to the box itself.

Once our new chassis was in the field, it was a success. After some time, we made a trip to see some customers who had purchased this new equipment. We observed something interesting at almost every customer's site. The equipment-users would go the chassis, pick it up, look at all sides, and then carry it over to their lab bench. Time and time again, we saw this in the field. A user would pick up a chassis, look at all the sides, and then carry it off. It made us wonder what was going on.

What we didn't realize was that the customers were searching for a handle; a simple little handle. As it turns out, the users of our equipment are accustomed to carrying similar devices throughout their labs. When we would use the equipment at our development lab it, for the most part, stayed put in one place, either in someone's cube or on a bench in the lab. We didn't realize the customer's mobility requirement. If we had beta tested this, and watched closely with a focus group mentality, we would have quickly seen that the user was grasping for a handle that didn't exist.

To keep The *Four Keys to Successful Design* from crumbling apart, we need to ensure that the highest of the four keys, Creation, does not fail. We've come too far to have a faulty layer crush our entire structure. As mentioned earlier, success breeds inspiration, a vital element among all the keys. If we can't get our innovations from thought to finish, and no one buys into them, we risk the chance of hitting a mass quantity of discouragement that could get in the way of our inspiration. As pointed out so many times throughout this book, we cannot risk losing the strength from any of the four keys. Stay focused on your goals with the help of your audience. Involve them early and often so you can shift course as needed, and not fall flat when it may be too late.

Listen closely to your audience, as they are the ultimate designers. Capture their attention early; otherwise, you may lose it later on. Keep your audience involved from Act 1, and they'll stick around for the grand finale.

17

Post Mortem

The last step during your excursion through the key of Creation is known as the Post Mortem, often referred to as a "Lessons Learned" exercise. It's through this exercise that you reflect back on your project, compile a list of the successes and failures encountered, prioritize them, and learn from them. From this prioritized list, you can then see what lessons need the most attention.

It's important that we realize our shortcomings and learn from them. The post mortem should never be a finger pointing exercise, nor should it be a time when you sense defeat by facing your failures. Think of this as a time to reflect on the good and bad of the project: what went well, and what didn't.

Mistakes help us build a better tomorrow. If we never face the reality that we all make mistakes, and that practically everything can be improved, then we'll never advance, and neither will our designs.

Many engineering students have learned one of the most historic events upon which many lessons were learned, the Tacoma Narrows Bridge disaster of 1940. You may have seen footage of this amazing disaster on television or in film documentaries. This was a massive bridge that was filmed being twisted and tossed about until it collapsed.

"Galloping Gertie," as this infamous bridge near Seattle, Washington, was also known, was regarded in its day as a marvel of engineering. It had the highest span-to-width ratio and span-to-depth ratio of any suspension bridge of its time.

Wind gusts through the Narrows would cause this bridge to oscillate, and it soon became a tourist attraction for the curious, who would watch it bend, sway, and contort. It was a disaster (literally) waiting to happen.

Then one day, November 7, 1940, at about 11:00 AM, the wind picked up to a higher than normal speed of 42 miles per hour through the Narrows. This caused greater vertical oscillations, and finally a cable connector slipped, and the bridge failed progressively, tearing itself apart.

There was only one reported fatality from the collapse of the Tacoma Narrows Bridge, a dog named Tubby, who was left in the only car stranded on the bridge. Leonard Coatsworth, a Tacoma newspaper editor who escaped from the stranded car, graphically provided an eyewitness account of "Galloping Gertie's" annihilation:

> "Just as I drove past the towers, the bridge began to sway violently from side to side. Before I realized it, the tilt became so violent that I lost control of the car…I jammed on the brakes and got out, only to be thrown onto my face against the curb.

> "Around me I could hear concrete cracking. I started to get my dog Tubby, but was thrown again before I could reach the car. The car itself began to slide from side to side of the roadway.

> "On hands and knees most of the time, I crawled 500 yards or more to the towers…My breath was coming in gasps; my knees were raw and bleeding, my hands bruised and swollen from gripping the concrete curb…Toward the last, I risked rising to my feet and running a few yards at a time…Safely back at the toll plaza, I saw the bridge in its final collapse and saw my car plunge into the Narrows."

Such a tragedy, although very sad, allows others to learn from its mistakes. How lucky it was that hundreds of people weren't killed when this 5,939 foot bridge collapsed. Fortunately, many engineers learned to build better bridges as a result of Gertie's disastrous demise.

The science of bridge aerodynamics was born after the Tacoma Narrows Bridge collapsed in 1940. The Tacoma experience taught engineers that wind causes not only static loads on the bridge, but also significant dynamic actions. They learned that a cable supported bridge is subject to wind-induced drag (a static component), flutter (the instability that occurred at Tacoma Narrows), and buffeting (where gusts "shake" the bridge).

Resolutions for future bridges included the use of wind tunnels and other forms of modeling to analyze these newfound effects.

All projects offer lessons that can be learned. There is no need to wait for your bridges to collapse to tackle these issues, either. Although the post mortem is typically done at the end of a project, feel free to do it at any time. Make this a fun exercise wherein everyone can feel comfortable. Bring in everyone related to the project, and have a good time with it. Perhaps order pizza or get together for lunch. Create an atmosphere where everyone can feel comfortable and address the lessons of the project without shame.

To help you with this exercise, I've provided an example spreadsheet shown in figure 8:

Lesson	Resolution	Importance	Likelihood	Priority
Chip vendor late with large quantities	Order earlier	5	6	30
Fine threaded holes in aluminum tended to strip	User new mill and bits	4	5	20
Using Java increased memory footprint	Use C++ for time critical components	8	7	56
Project went into summer and more people were on vacation	Allot time in schedule	3	5	15
Latex material tended to melt due to friction	Use kevlar at friction points	8	8	64

Figure 8 Lessons Learned Example Tool

This tool is similar to the issue management spreadsheet we used earlier in Chapter 11, "Find Your Obstacles Early and Often," with some slight variations. The first column lists each lesson that was learned. The second column lists the resolutions for the lesson to avoid it next time. The remaining columns are used to bring these lessons into a prioritized view so you can manage them accordingly.

To use the "Lessons Learned" tool, before you meet, have everyone involved with your project think of all the lessons that were learned. Have each person fill in the first column with the lessons they can think of. You, as the moderator, should take everyone's individual list of lessons and compile them into one spreadsheet.

Then get everyone together to discuss the lessons. If the lesson was something that didn't go well during the project, think of what you could do differently and list these "resolutions" in the Resolution column. Bear in mind that not all lessons are things that didn't go well. Some lessons may have turned out okay or gone extremely well. In this case, you want to make sure you repeat these good lessons the next time, so feel free to list them and their resolutions as well.

Next, as a group, come to a decision on a weight from 1 to 10 to rank how important each lesson is to future projects. Use 1 for the lowest importance, and 10 as the most important, and place this in the Importance column for each lesson. Then give a similar weight between 1 and 10 to the "likelihood" of this issue recurring, and place this in the Likelihood column.

The last column, Priority, should be an automatic calculation of "Importance" multiplied by "Likelihood," resulting in the priority level at which this lesson needs to be addressed. As with the issue management tool we used in Chapter 11, keep this spreadsheet sorted by priority with the highest priorities first.

After your post mortem group discussion, archive this tool so you can reference it for your next project and innovation. Don't let this data gather dust. Before you start your next innovation, break out this tool and review it. Keeping these lessons in the forefront of your mind will expedite your next innovation, resulting in a higher quality product, which in turn increases your inspiration.

Never be afraid of lessons you can learn from mistakes or mishaps. These are not things to be shunned. Instead, they should be welcomed as the foundation to build better things. Next time you drive over a bridge, remember how safe it is due in large part to lessons learned from the Tacoma Narrows incident. And each time you turn on a light, remind yourself of Edison's 3,000 attempts to make a successful bulb. We all make mistakes. It's part of human nature and the design process. However, most importantly, we need to learn from our mistakes.

To close out this chapter on the last of the four keys, I'd like to leave you with some parting thoughts, spoken wisely from some great people:

The only real mistake is the one from which we learn nothing.—John Powell

A life spent making mistakes is not only more honorable, but more useful than a life spent doing nothing.—George Bernard Shaw

We ought not to look back unless it is to derive useful lessons from past errors, and for the purpose of profiting by dear-brought experience.—George Washington

It is not the critic who counts, not the man who points out how the strong man stumbled, or where the doer of deeds could have done better. The credit belongs to the man who is actually in the arena, whose face is marred by dust and sweat and blood, who strives valiantly, who errs and comes short again and again, who knows the great enthusiasms, the great devotions, and spends himself in a worthy cause, who at best knows achievement and who at the worst if he fails at least fails while daring greatly so that his place shall never be with those cold and timid souls who know neither victory nor defeat.—Theodore Roosevelt

Summation

We used to think that if we knew one, we knew two, because one and one are two. We are finding that we must learn a great deal more about 'and'.

—Sir Arthur Eddington

Congratulations! You now have in your possession the four keys to unlock your potential for successful design!

As pointed out so often in this book, these keys are truly intertwined and dependent on each other. If it not for one, the others would have little meaning. It's safe to say that it is the sum of the parts that is equal to, yet sometimes greater than, the whole. We need a healthy, equally divided formula of Inspiration, Innovation, Exploration, and Creation to produce a successful design.

If I could put this in geeky math terms, the concept might look something like this:

$$\sum_{i=1}^{4} X_i = Successful\_Design$$

Where:

- X_1 = Inspiration
- X_2 = Innovation

- X_3 = Exploration
- X_4 = Creation

Adding all these keys together, we achieve our ultimate goal: a successful design.

But let's leave that geeky math stuff for the office. My brain still hurts from making that formula. Instead, let's quickly "sum up" what we've learned from these four keys in more simple terms.

To design successfully, we need inspiration. This is the bottom rung on the food chain comprising the four keys. Without a continuous diet of inspiration, our innovation will suffer, as will the breadth of our exploration, resulting in a poor creation or no creation at all.

Success breeds inspiration, but learn to utilize failures to do the same. If it were not for a failure in the 3M lab one day in 1970, we wouldn't have the Post-it® Notes we use today. Failure inspires success. Just keep your eyes open and learn to look at failures differently. See them not as defeats, but opportunities.

To stay inspired, secure the proper environment. Don't waste your years working in an environment that cannot motivate you and feed your inspirational appetite. Remember that your inspiration and innovation are too precious to waste. Nurture your inner engineering essence by maintaining a positive work environment. Inspiration will be less inhibited, resulting in more powerful innovative thought.

Look inside yourself to see what inspires you. Find those things you are good at and that come naturally to you. Pursue those angles as possible paths in your career. Doing so will help you build your inspiration in a natural way. We all have inspiration. Sometimes we just need to find out what inspires us the most by looking deep inside ourselves, asking some pointed and difficult questions, and answering them honestly and truthfully.

Innovation is not just doing, it's thinking—constantly thinking. Always have wonder, and be constantly curious. If Charles Goodyear hadn't been in the right frame of mind, one of inspirationally driven innovative thought, he could have just scraped off some spilled rubber and sulfur, thrown it away, and gone on to something else. Had he done that, how long would it have been until we had vulcanization to revolutionize the way car tires are made today?

Innovation is finding a deep hole called "need" then filling it with a solution. Necessity is not only the mother of invention; she births innovation as well. Being inspired, our innovation is like a tinderbox; one slight spark of an idea can ignite a wildfire of innovations.

Use discouragement directed at your innovations as constructive input. Learn to sort the positive aspects of all feedback to your advantage. Discouraging events do not directly hinder your mood and inspiration. There's a middleman in your brain, your internal dialogue. Once realized, you can manipulate this intermediary component to your benefit, filtering out the negative aspects of discouragement while maintaining the sweet juicy center of positive critique.

As you innovate, step off your island and visit the land outside your immediate circle. Gather input on your innovations from various sources, and thrive on versatility. Don't get caught up in a next-bench approach. Instead, invite the ideas and critiques of various sources, thereby increasing your innovation's versatility.

Don't reinvent the wheel unless you have to. Innovation is quite often a way to make things better. Henry Ford didn't invent the car; he produced cars faster. Ford built a better mousetrap that revolutionized the automotive industry. Look for ways to improve rather than starting everything all over from scratch.

Explore all possibilities. Don't be afraid of failure. Be persistent and don't give up. Edison made 3,000 attempted theories for his filament bulb before he got one that worked well. If not for his persistence, would we have the light bulbs we know of today? We build our tomorrow on the failures of today.

Find your obstacles early as you explore; do not wait for them to find you. Find as many obstacles and issues that you can. Don't be afraid of obstacles. Let them know you're the biggest dog on the block, and you can take them all on, any day, anywhere. Use an issue management tool like the one in Chapter 11 to get a good grip on the issues. If you're working in a group, foster a culture in your teams, and think of rewards for those that find the problems, not just those who solve them.

Learn from the mistakes others have made, as well as from your own. Exploration is a matter of knowing what won't work, as well as what will. You will make mistakes; it's human nature. Just make sure you take the time to learn from them.

Learn to iterate through the key of Creation, slowly molding your idea into fruition, and staying inspired as you watch it progress. Creation is not a one-step process. Like a tree that grows, Creation is a systemic entity that sprouts from the seed of innovation and exploration, then slowly blossoms through an iterative progression. As this product of your idea grows through each iteration, you maintain perspective and continue to feed your inspiration. This in turn feeds your innovation, which feeds your desire to explore. Iteratively stepping through the key of Creation requires dependency on the other three keys.

Build what the customer wants, period. Don't go overboard on your design unless you have to. Keep things effective and in budget, and get them to market in a timely manner. If the customer asks for a glass of milk, don't design him a dairy farm with flying cows.

As you make something, keep the audience in view. Listen to what they have to say, and remember: they are your ultimate designers. Think of focus groups, or similar activities in which to involve your customers during the key of creation. If you show your audience Act 1, they'll likely stick around for the grand finale.

When it's all done and said, it's not all done and said. Go back over the entire project and learn from the things that went well, and those that didn't. Involve those who participated in the project, and perform a post mortem exercise, capturing the lessons to be learned for future innovations.

Don't give up, and never throw in the towel.

Be ready for the next adventure!

Wash, rinse, repeat…

Afterwords

You gain strength, courage and confidence by every experience in which you really stop to look fear in the face. You are able to say to yourself, 'I have lived through this horror. I can take the next thing that comes along.' You must do the thing you think you cannot do.

—Eleanor Roosevelt

The four keys presented in this book may unlock your potential in creative and productive design. Nevertheless, they only scratch the surface. To drill deep into the various aspects presented throughout each of the chapters, I might have filled this book with tedium. Doing so, I would have risked enhancing many areas that may not be of interest to you.

I wrote this book using a "broad and shallow" approach, covering a broad range of topics, but only wading into the shallow end of each one. I encourage you to dive into the deep end of points where you feel you need further information. To help you find more answers, and advance your mastery of the four keys, I've included a "Recommended Reading" section following my parting thoughts.

I also have plans to provide other similar works to tunnel deeper into the world of inspired and creative design. To help foster a community where you can contribute your thoughts on this book, good or bad, and to bring together our collective thoughts, please feel free to visit The Four Keys web site at:

http://cool-net.com/4keys

You can also contact me directly by e-mailing me at: 4keys@wavecast.com

Together we truly have collective thought, and power in numbers. Each of us being unique not only varies our inspirational appetites and innovative thought, but also the further knowledge we need on specific subject matters.

Many people have said "…I could write a book about it…" when referring to adversity they have faced in their lives. I said the same thing, but also kept track of the great things in life that inspired me, and sparked my innovation. Through years of exploration, and observations to bring things to creation, I felt my collection of experiences and influences could be best shared with others, such as you. I've seen great things come to fruition, and have seen many more (not listed in this book) fall to utter failure. To this day, I still at times have to be an innocent bystander, painfully watching others around me build gingerbread houses on anthills. I cannot teach everyone my experiences, but hopefully others will learn well from their own. In addition, I can only hope that I learn from the experience of others as well.

In explaining the four keys, I wanted to share with you those things I've found to work, and not work, in my career. Many references to the greats, like Edison, Einstein, the 3M story of Post-It® Notes, and others, reflect similar lessons I've learned throughout my career. As I convey to you my desire to share my knowledge for your education and insight, I also encourage you to learn from the historical figures that influenced me; many of which I referenced throughout this book. Once we think that we cannot learn from those who have paved the way for us, we lose the foundation that our world is built upon: the inspirations, innovations, explorations, and creations of the great men and women before us. Don't throw away the past; seek it, savor it, and learn from it.

Lastly, with the utmost sincerity, may I say that I truly wish the four keys help you throughout your career, whatever it may be. I truly hope that all your successes, as well as your failures, bring you the highest level of gratification in the knowledge that you have pursued your dreams.

Stay inspired, be innovative, explore this world, and create things for the good of humankind.

Nathan Todd Cool

Recommended Reading

How many a man has dated a new era in his life from the reading of a book.

—Henry David Thoreau

The lessons I've learned throughout life came not only from experience, but from the works of many famous authors as well. If not for the great authors of books and articles I've read, I would not have as many insights to share. Therefore, I want to share with you some books that I find useful, and used as references throughout "The Four Keys to Successful Design."

For further information on finding your strengths, skills, and talents, I recommend "Now, Discover Your Strengths" written by Marcus Buckingham and Donald O. Clifton. This book elaborates on points raised in the Inspiration key, and helps you to understand how and why you have the skills you do. An additional online survey assists in the strength finding exercise.

Additional information for strength finding, as well as other motivational topics, including topics surrounding "Human Givens" therapy, can be found online at Jane Firbank's web site:

http://www.janefirbank.com/

For dealing with cognitive distortions, and other aspects of mood perception, I highly recommend "Feeling Good, The New Mood Therapy" by David D. Burns, M.D. This book sheds some amazing light on how we perceive things, how the mind works, and how to tackle various problems, from depression to anxiety.

Also on the subject of cognitive distortions, I recommend some online articles written by Nancy Schimelphfening that are posted at:

http://depression.about.com/.

Homer Hickam, the original "rocket boy" has written a variety of books, including "The Rocket Boys," "The Coalwood Way," and "We Are Not Afraid." Homer's experience and inspiration comes shining through in his work. You can find more on Mr. Hickam's books on this web site:

http://www.homerhickam.com

One book that helped me realize the human spirit of innovation and evolution of man is "Guns, Germs, and Steel—The Fates of Human Societies" by Jared Diamond. This book gives a phenomenal perspective on how we came to be, not necessarily via the genetic focus of evolution, but how the inspiration, innovation, and exploration of humankind has led us to where we are today.

For more information on the Unified Process (UP), I recommend two books. The first is the basics of UP, "The Unified Process Explained" by Kendall Scott. The second book I recommend is more geared toward Object Oriented Programming, and is "Applying UML and Patterns, An Introduction to Object Oriented Analysis and Design and the Unified Process" by Craig Larman. This book, despite its long title, is very informative, shows derivations of the UP, and how to apply it to OO programming.

When listening to your audience, I recommend books on focus groups such as "Focus Groups: A Practical Guide for Applied Research" by Richard A. Krueger and Mary Anne Casey. Richard A. Krueger also has another useful book on this subject, "Developing Questions for Focus Groups."

To improve your personal experiences in life and in the workplace, I recommend "The 7 Habits of Highly Effective People" by Stephen R Covey. This popular book explains how we can take some simple principles and apply them into our lives, thereby being more effective and comfortable.

Acknowledgements and References

This book would not have been possible if not for the valuable resources that assisted in my research. I would like to thank all the authors of the following books, web sites, and articles for their efforts in creating resources for others to learn from:

Quotations: Many quotations were used from "The Quotations Page" at:
http://www.quotationspage.com/subjects/
Special thanks to Mr. Moncur for his site and permission for use.

Various history references were researched through the "Hyper History" web site at:
http://www.hyperhistory.com/

The story of Velcro® was researched on the Velcro® web site at:
http://www.velcro.com/

Various invention stories were researched using the following sites:
About.com at http://inventors.about.com
Idea Finder at http://ideafinder.com
Special thanks go out to Mary Bellis from Inventors.About.Com for her permission in fair use, and for putting together such a fact-filled web site.

Computer World's online report "The Best Places to Work in IT" (2002), referred to in Chapter 3 (The Company You Keep), is publicly available at:
http://www.computerworld.com/bestplaces2002/

The Rice University employee satisfaction survey, referred to in Chapter 3 (The Company You Keep) is publicly available on the web at:
http://facilities.rice.edu/studies/survey2002.pdf

"Guns, Germs, and Steel, the Fates of Human Societies" by Jared Diamond was used as a reference for research when referring to Polynesian societies and sociological development (i.e., in the Next Bench chapter).

Patent abstracts, descriptions, and diagrams were found and retrieved in accordance with public domain Copyright law (17 U.S.C. § 105) from the United States Patent and Trademark Office at:
http://www.uspto.gov/

Cognitive distortion was researched using two primary sources: "Feeling Good" by David D. Burns, M.D., and online articles by Nancy Schimelphfening, which are posted at:
http://depression.about.com/
I'd like to extend a special thanks to Nancy Schimelphfening for allowing me to paraphrase portions of her article on cognitive distortions, and to Dr. Burns for his positive response and inspiration.

Strength finding excerpts in the Introspection chapter (4) were used by permission from Jane Firbank from her articles at:
http://www.janefirbank.com/
In particular, the following article was excerpted and paraphrased for context:
http://www.janefirbank.com/articles/strengths.html

Homer Hickam's story and biographical data can be found at his web site:
http://www.homerhickam.com/
With permission from Homer Hickam, every attempt was made to keep the story line in the "To The Moon" chapter (13) as accurate as possible to Homer's book, "The Rocket Boys," and not the movie, which has some inaccuracies. I'd like to extend a special thanks to Homer for his permission and continued inspiration.

3M references to Post-It® Notes history, as well as William L. McKnight's culture statement, were researched at the 3M web site:
http://www.3m.com
Special thanks to Mr. Germain and consumer relations at 3M for your assistance and the permission to paraphrase the famous Post-It® Notes story of success.

Facts and dates for the Tacoma Narrows disaster discussed in Chapter 17 ("Post Mortem") were researched at the following web sites:
http://www.civ.toronto.edu/funstuff/disaster/tacoma.htm
http://www.cen.bris.ac.uk/civil/staff/dib/casehist99/tacoma.htm

Partial wording for the "Next Bench Marketing" description was used by permission from Reed Electronics Group's "Electronic News" article by Steve Leibson (6/25/2001) located at:
http://www.reed-electronics.com/electronicnews/index.
asp?layout=article&articleid=CA90459
Special thanks to Mr. Sperling for your permission for the partial use and for allowing me to paraphrase for context.

Famous quotes relating to Charles Duell, Lord Kelvin, Marechal Foch, Thomas Watson, and Steve Jobs were researched using:
http://www.goofups.com/
Special thanks to Anup, owner of goofups.com

Story of Apple® computers in Chapter 7 ("Expansion from Discouragement") was researched online from the following sources:
http://www.apple.com/pr/bios/jobs.html
http://www.ideafinder.com/history/inventors/jobs.htm

Verse on straining gnats in Chapter 11 is from the Bible, Mathew 23:24.

0-595-29909-1